Tumor Therapy with Tumor Cells and Neuraminidase

Contributions to Oncology
Beiträge zur Onkologie

Vol. 27

Series Editors
S. Eckhardt, Budapest; *J. H. Holzner,* Wien;
G. A. Nagel, Göttingen

Basel · München · Paris · London · New York · New Delhi · Singapore · Tokyo · Sydney

Tumor Therapy with Tumor Cells and Neuraminidase

Cell Physiological, Immunological, and Oncological Aspects

H. H. Sedlacek, Marburg

19 tables, 1987

Basel · München · Paris · London · New York · New Delhi · Singapore · Tokyo · Sydney

Contributions to Oncology
Beiträge zur Onkologie

Drug Dosage

    The authors and the publisher have exerted every effort to ensure that drug selection and dosage set forth in this text are in accord with current recommendations and practice at the time of publication. However, in view of ongoing research, changes in government regulations, and the constant flow of information relating to drug therapy and drug reactions, the reader is urged to check the package insert for each drug for any change in indications and dosage and for added warnings and precautions. This is particularly important when the recommended agent is a new and/or infrequently employed drug.

All rights reserved.

    No part of this publication may be translated into other languages, reproduced or utilized in any form or by any means, electronic or mechanical, including photocopying, recording, microcopying, or by any information storage and retrieval system, without permission in writing from the publisher.

© Copyright 1987 by S. Karger GmbH, Postfach 1724, D-8034 Germering/München and
S. Karger AG, Postfach, CH-4009 Basel
Printed in Germany by Gebr. Parcus KG, München
ISBN 3-8055-4549-5

Contents

Preface	VI
Abbreviations	VII
Introduction	1
Changes in Cell Surface	3
The Effect on Cell-to-Cell Contact and General Properties of the Cell Membrane	3
Influence on Membrane Receptors	6
Alterations in Cell-Specific Antigen Structure	8
Demasking of Neoantigens (Thomsen-Friedenreich Antigen and Immune Response Antigens)	12
The Appearance of Extraneous Antigens on Cells	15
The Influence on the Immune System	21
Influence on Cell Immunogenicity	21
Adjuvant Activity of VCN	23
Influence on Granulocytes, Macrophages, and Lymphocytes	27
Adjuvant Effect for Tumor Cells	32
Experimental Tumor Therapy Tests	35
Administration of VCN-Treated Inactivated Tumor Cells	35
Combination of VCN-Treated Tumor Cells and Non-Specific Immunostimulants	42
Administration of Mixtures of Tumor Cells and VCN	43
Chessboard Vaccination	45
The Role of Thomsen-Friedenreich Antigen	53
Clinical Studies	56
VCN-Treated Tumor Cells	56
Combination of Tumor Cells and VCN	61
Summary	67
Acknowledgments	69
References	70

Preface

Vibrio cholerae neuraminidase (VCN) is known to have a broad specificity in cleaving sialic acid from glycoproteins and glycolipids. As sialic acid is a strongly negatively-charged sugar moiety, cleavage of sialic acid from the cell surface by VCN was thought to increase the immunogenicity of those cells by unmasking or exposing membrane antigens. Based on this hypothesis, VCN-treated tumor cells have been used for specific immunotherapy of cancer, both experimentally as well as in clinical studies. The results so far, however, have revealed either unreproducibility of the therapeutic effect or even tumor growth enhancement induced by this kind of treatment.

Initiated by the finding that VCN very strongly attaches to the membrane of VCN-treated cells, the role of VCN as an adjuvant for the immune response has been elaborated. Moreover, new vaccination procedures were developed to overcome the problem of tumor enhancement induced by the application of an inadequate number of VCN-treated tumor cells.

These efforts led to the development of the 'chessboard vaccination', that is, the simultaneous intradermal application of increasing numbers of tumor cells, each mixed with increasing amounts of VCN. This chessboard vaccination was found to be superior to all other kinds of specific tumor immunotherapy with the use of VCN. An increase in remission duration or survival by chessboard vaccination could be found in the therapy of metastasizing transplantation tumors (Lewis lung AC) in mice, of spontaneous mammary tumors in dogs, and even of colon carcinoma (Duke C) in man.

These clinical data have to be confirmed and widened, of course. Nevertheless, they induce hope that a specific tumor immunotherapy might become an effective treatment in selected tumor diseases, in spite of the fact that this hope is more than 10 years old and has suffered from numerous set-backs in the past.

Marburg, January 1987 *H. H. Sedlacek*

Abbreviations

AML	acute myeloid leukemia
ARBC	autologous red blood cells
BCG	bacillus Calmette-Guérin
CFA	complete Freund's adjuvant
ConA	Concanavalin A
CPN	Clostridium perfringens neuraminidase
CPPD	Cyclophosphamide
DTH	delayed type hypersensitivity
IFA	incomplete Freund's adjuvants
IVN	influenza virus neuraminidase
LAF	lymphocyte-activating factor
MCCNU	Methylchlorethylcyclohexylnitrosourea
MER	methanol-extracted residue
MHC	mean histocompatibility complex
MLC	mixed lymphocyte culture reaction
NeuAc	N-acetyl neuraminic acid
PBS	phosphate-buffered saline
PFC	plaque-forming cytolysis
PHA	phyt-hemagglutinin
PPD	purified protein derivative of mycobacterium
SRBC	sheep red blood cells
TAA	tumor-associated antigens
VCN	Vibrio cholerae neuraminidase
WGA	wheat-germ agglutination

Introduction

Enzymes which split the ketosidic binding between neuraminic acids and their partner molecules [41, 124, 202] are called neuraminidases (E.C.3.2.1.18: Enzyme nomenclature 1965).

Various neuraminidases have been found in myxoviruses [86], various bacteria, protozoa [238], and vertebrate organs [114, 126]. These neuraminidases differ in their substrate specificity [87, 443]. The neuraminidase of Vibrio cholerae (VCN), for instance, cleaves $2-3$, $2-6$, and $2-8$ glycosidic linkages at different speeds of reaction, whereas the neuraminidase of Newcastle disease virus only hydrolases the $2-3$ and $2-8$ glycosidic bindings [85, 86, 289, 443]. Furthermore, the neuraminidases of various origins differ in molecular weight, pH-optimum, Ca^{++}-dependency, and Michaelis constant [238].

Whereas the structure of influenza virus neuraminidase has already been elaborated in detail [57, 442], corresponding data for bacterial neuraminidases do not yet exist. The following characteristics are known for VCN: molecular weight: 90,000 daltons [4], 10,000 daltons [237], and 68,000 daltons [86]; isoelectric point: 4.8 [243]; pH-optimum: 4.5 [4] and 5.5 [334]; Michaelis constant (K_m) in the range of 0.9×10^{-3} to 7×10^{-3} mol/l [231, 425]; a Michaelis constant of 1.2×10^{-3} has been found for the substrate sialyl-lactose [4, 86].

The enzymatic activity of VCN is dependent on calcium ions (Ca^{++}) [4, 44]. Nothing is known of the carbohydrate part or the amino acid composition of VCN.

Experiments using Vibrio-cholerae-neuraminidase-treated tumor cells for immunotherapy of tumors have their origin in the assumption of Ruhenstroth-Bauer et al. [324] that NeuAc may play a decisive role in tumor genesis: An increase of NeuAc on the membranes of cells enhances their negative charge and their mutually repelling forces and may correlate with the degree of malignancy of tumor cells. This final hypothetical conclusion has not been substantiated by experimental evidence [60, 90, 128, 224, 225, 256, 392, 454]. It can be assumed, however, that it was the basic idea behind the aim of increasing the immunogenicity of normal cells and of tumor cells by cleaving membrane-bound N-acetyl neuraminic acid (NeuAc). In the main, Vibrio cholerae neuraminidase (VCN) was used for splitting off NeuAc because it has a broad substrate specificity and because it is an inducible enzyme [335] which can be isolated out of the culture broth of Vibrio cholerae to a high pureness [334]. Thus, in the preparation commonly used for biological studies, neither proteases, phospholipase C, N-acetyl neuraminic acid aldolases, nor endotoxins could be detected [310, 335].

Changes in Cell Surface

The Effect on Cell-to-Cell Contact and General Properties of the Cell Membrane

The treatment of cells with VCN leads to the separation of terminal N-acetyl neuraminic acid (NeuAc), mainly from the membrane-bound carbohydrate arms of the glycoproteins [3, 55, 59, 460] which are probably arranged in a 'cluster-like' formation [454], and, to a lesser extent, from the glycolipids in the outer lipid layer [26, 455, 467]. Concomitant with the separation of NeuAc, the negative surface charge of the cell is diminished since NeuAc is the cause of the major part of the negative cell-surface charge [323, 451]. These changes in cell-surface charge lead to a diminution of the anodic electrophoretic mobility of the cell [108, 230, 323, 326, 393, 412, 451, 452] and go hand-in-hand with an increased cell deformability [451, 453], an enhanced cell adherence [61, 291] and cell aggregation [74, 174, 332]. In all probability, these phenomena are brought about by a reduction of the NeuAc-mediated repulsion forces between the cells [454].

A simple model for measuring cell aggregation is the spontaneous rosette formation of various erythrocytes to lymphocytes from donors not immunized with erythrocytes.

The so-called spontaneous rosette formation of lymphocytes is most marked with sheep red blood cells (SRBC) [21, 39]. It is, however, not limited to SRBC but can be demonstrated for erythrocytes of other species as well [367]. In particular, rosette formation with SRBC is used as a marker for thymus-dependent lymphocytes [106, 173, 206, 462]. The percentage of rosettes with SRBC is drastically raised after VCN treatment of lymphocytes and/or erythrocytes [35, 81, 110, 115, 340, 367, 369, 450, 466]. Moreover, rosettes even with homologous or autologous erythrocytes (table I) could be found by us after VCN treatment of lymphocytes and/or erythrocytes [369], and the lymphocytes involved have also been attributed to the T-cells [27, 56, 142].

Table I. Influence of VCN on the number of rosettes spontaneously formed between human pheripheral lymphocytes and erythrocytes

Lymphocytes		Rosettes with	
pretreatment	subsequent treatment	SRBC	ARBC
—	—	52.4 ± 5.3[a]	0
VCN	—	74.8 ± 1.2	43.8 ± 10.1
VCN	4°/20 h	71.0[b]	39.0
VCN	37°/20 h	53.0	8.0
—	Fc	51.0	0
VCN	Fc	69.0	35.0
—	asialo-G	56.0	0
—	monosialo-G	18.0	0
—	disialo-G	13.0	0
—	trisialo-G	0	0
VCN	asialo-G	80.0	32.0
VCN	monosialo-G	32.0	7.0
VCN	disialo-G	30.0	1.0
VCN	trisialo-G	12.0	0

VCN: 100 mM + 7.5 × 10⁶ lymphocytes, 15 min, 20°C; subsequently 3 washes
Fc: 5 × 10⁶ lymphocytes + 0.1 ml Fc/IgG (1 mg/ml); 60 min, 20°C; subsequently 3 washes
Gangliosides: 5 × 10⁶ lymphocytes + 1 ml gangliosides (100 mM), 60 min, 37°C; subsequently 3 washes
Asialo-G = G_{tet}-Cer; monosialo-G = G_{Gtet}-1; disialo-G = G_{Gtet}-2b;
Trisialo-G = G_{tet}-G
SRBC: sheep red blood cells
ARBC: (human) autologous red blood cells
[a] mean ± s of 10 donors
[b] one of 3 identical experiments
According to Seiler and Sedlacek [369]; Seiler et al. [370]

The basic mechanisms and structures underlying rosette formation prior to or after VCN treatment of lymphocytes or erythrocytes are unknown. However, sialic acid (NeuAc) on the cell membrane appears to be involved, as T-cells possess considerably more membrane-bound, VCN-cleavable NeuAc than B-cells or thymocytes [78, 300, 393, 412], whereby NeuAc appears to be an essential component of the Thy-1 antigen [152, 445]. T-helper cells and T-suppressor cells differ further in the

NeuAc and galactose structures of membrane glycoproteins and glycolipids [240].

After cultivation of VCN-treated lymphocytes over 20 h at 37°C (not, however, at 4°C), the VCN-mediated increase in rosettes with SRBC or the appearance of rosettes with autologous erythrocytes is reversible, probably due to the temperature-dependent resynthesis of NeuAc-containing, membrane-bound glycoproteins [369]. This assumption is supported by experiments with hamster ovarial cells, where, after treatment with VCN, the resynthesis of membrane-bound NeuAc is completed under culture conditions (37°C) within 12–16 h, a process which can be inhibited by the inhibition of protein synthesis with puromycin [199].

Spontaneous rosette formation can also be inhibited by absorption of purified gangliosides into the cell membrane of lymphocytes, demonstrated in our experiments by means of radioactive gangliosides [370]. This inhibition of both rosette formation with SRBC and of that with autologous erythrocytes was clearly dependent on the neuraminic acid content of the gangliosides used (table I). Whereas asialo-ganglioside was seen to be ineffective, the sialo-gangliosides, especially trisialoganglioside, inhibited the rosette formation of VCN as well as of untreated lymphocytes. Altogether, these experiments point to the probability that the increase in the number of rosettes after pretreatment of erythrocytes and lymphocytes with VCN is at least partially caused by a decrease in membrane-bound NeuAc and, thus, of mutually repelling negative charges. Additional factors might be lectin-like receptors on lymphocyte cell membranes (similar to those on liver cells [193–195] or Kupffer cells [13] or on macrophages [201]) for galactose-containing membrane structures, which are more strongly exposed after cleavage of NeuAc from the surface of the cell membrane. Another interpretation would be that structures underlying spontaneous rosette formation without VCN influence differ from those after VCN treatment.

Since rosette formation, with SRBC as well as with VCN-treated autologous erythrocytes, can be inhibited by means of cytochalasin-B treatment of lymphocytes, microfilaments probably participate in both systems [56].

In a way similar to the enhancement of rosette formation after VCN treatment, the VCN-induced enhancement of macrophage phagocytosis [190, 207, 454] – even in the presence of opsonizing factors [338, 339] – and the increase in macrophage cytotoxicity [374] or of the NK-cell activity [456] after VCN treatment of the relevant target cell may at least partially be caused by an alteration in cell surface charge. It is improbable that a direct cytotoxic effect of VCN on the target cell is the cause, in particular, of increased macrophage cytotoxicity since, in the VCN concentration used, treated cells proliferate both in culture and in vivo after transplantation [158], protein synthesis is not influenced [313, 314], and there is no impairment of the cell membrane barrier − measured by dye exclusion (trypan blue) [299, 315, 317, 329, 461].

However, the resistance of the cell membrane does seem to be reduced, since VCN-treated cells (unlike cells incubated with influenza virus neuraminidase) possess a higher degree of sensitivity [165, 284, 285, 287] for complement-dependent cytolysis induced by xeno- or allo-antibodies or by cobra venom factor. A reduction of repulsion forces, however, may also be significant here [70].

Comparative studies with soluble and carrier-bound VCN testify against the assumption that the increase in macrophage phagocytosis of VCN-treated target cells is solely caused by a reduction in repulsion forces. Whereas target-cell treatment with soluble VCN caused strong phagocytosis, the separation of the same amounts of NeuAc by carrier-bound VCN had only marginal effects [252].

Further phenomena which can, at least superficially, be explained by a reduction of the negative surface charge are, for thrombocytes after VCN treatment, an increase in aggregation by ADP, serotonin, or collagen [72, 127] and, in vivo, enhanced clearance [127] and, after VCN treatment of endothelial cells, an increased granulocyte adhesion [122].

Altogether, it can be concluded that treatment with VCN induces a change in the physiochemical properties of cells − i.e. reduction in negatively charged repulsive forces, which appears to be an important reason for an alteration in behavior and function of cells.

Influence on Membrane Receptors

Investigations on receptors for the Fc-part of IgG on the cell membrane of lymphocytes (B-cells, T-cells, NK-cells, and K-cells) [2, 43, 206, 255, 259, 376, 459] served us as a model for answering the question as to whether VCN, through separation of NeuAc, is capable of demasking membrane receptors.

Fc-receptors were demonstrated with the aid of immune complexes comprising erythrocytes as antigen and a rabbit anti MN-antiserum (amboceptor) [369]. This immune complex binds to Fc-receptors on lymphocytes via the Fc-part of the antibody involved while forming rosettes [206]. The number of rosettes with erythrocyte-amboceptor complexes (EA-complexes) increased considerably after VCN treatment of lymphocytes [131, 163, 206, 344, 369]. The increase was reversible if the lymphocytes were cultivated at 37°C over 20 h, but not at culture temperatures of 4°C.

Both IgG and Fc-fragments of IgG, when added to the mixture of lymphocytes and erythrocyte amboceptor, caused a complete blockage of Fc-receptors, which means an inhibition of rosette formation with EA complexes without impairing spontaneous rosette formation with SRBC. Increased exposure of Fc-receptors after VCN treatment of lymphocytes, too, could be inhibited to a high degree with Fc-fragments

Table II. VCN-induced increase of human lymphocytes exposing membrane Fc-receptors

Pretreatment		% Lymphocytes, binding				
VCN (mU)[a]	Fc (IgG) (mg)[b]	rabbit anti-K-antibodies[c]		rabbit IgG[c, d]		erythrocyte-amboceptor-complexes[d]
		1:1	1:4	1:1	1:4	
—	—	35	18	8	6	28
—	1	34	20	3	4	0
—	0.01	28	19	1	1	0
100	—	87	73	55	34	58
100	1	19	17	5	4	7
100	0.01	38	18	5	3	50
100	— (4°/20 h)[e]	75		n.d.		56
100	— (37°/20 h)	39		n.d.		33

[a] 5×10^6 lymphocytes + 0.1 ml (100 mU) VCN; 20°C; 30 min, subsequently 3 washes
[b] 5×10^6 lymphocytes + 0.1 ml Fc-solution; 20°C; 30 min, subsequently 3 washes
[c] Evaluation by double antibody immunofluorescence method
[d] Detection of Fc-receptors
[e] Incubation after pretreatment
According to Seiler and Sedlacek [369]; Sedlacek et al. [353]

(table II), and to a lesser degree with IgG [369]. At the same time, increased spontaneous rosette formation with SRBC could be partially inhibited by high concentrations of Fc-fragments only. These specificity tests give weight to the assumption that Fc-receptors are more easily accessible after VCN treatment, i. e., after separation of membrane-bound NeuAc, which finally causes an increase in the number of lymphocytes forming rosettes with EA complexes.

It is questionable whether VCN treatment only demasks receptors for the Fc-fragment of IgG or also the receptors for IgM. On the one side, an increase of Fc-receptors for IgM both on lymphocytes [131] and macrophages [132] was observed, and, on the other, a complete destruction [163, 344] after cleavage of NeuAc.

These contradictions may result from the fact that Fc-receptors have so far been defined only functionally by determination of the saturation, specificity, and binding affinity of immune complexes to the cell membrane [84]; moreover, the frequency of their appearance is variable and dependent on external factors such as virus infection [229] or stimulation for instance of T-cells [185, 347] or macrophages [301].

It is unclear whether or not increase in Fc-receptors for IgG after treatment with VCN is related to an increase in functional capacity. Thus, after VCN treatment of lymphocytes, ADCC was either unchanged [344] or enhanced [444].

VCN treatment not only leads to an enhanced exposure of Fc-receptors on lymphocytes, but also to an increased sensitivity of T-cells for mediators produced by macrophages [188, 189]. There are also indications that after VCN treatment bone marrow stem cells become more susceptible to erythropoietin [273, 322] and that VCN treatment reduces the insulin-dependent glucose transport in fat cells, possibly due to an increase in signal transmittance after insulin-binding to its receptors [65].

Altogether, it can be concluded that cleavage of NeuAc by VCN from the cell membrane may increase the exposure and/or functional capacity of cell membrane receptors.

Alterations in Cell-Specific Antigen Structure

The finding that various tumor cells exhibit a higher VCN-cleavable negative surface charge than the corresponding normal cells [12, 73, 109] primarily led to the assumption that malignancy was associated with a rise in NeuAc-induced negative cell-surface charge. Several investigations were indeed able to detect certain associations between high negative surface charge, cell proliferation and maturity [90], mitosis [224], and growth rate [225], but not, however, with malignancy itself [60, 128, 256, 392, 454].

Logically, then, further studies took up the question as to whether NeuAc, if not associated with malignancy, could have any influence on the growth and recognition of normal cells and tumor cells, for instance by covering either naturally occurring (common antigens) or new, normally concealed membrane antigens [329]. Another aspect of the question was to what extent membrane-bound NeuAc was involved in or associated with the metastasizing tendency of tumors by masking terminal carbohydrates [431, 432]. Evidence from animal experiments has shown clear indications [11, 101, 102, 464, 465] that the metastasizing rate of a number of mouse tumor cell lines correlates positively with the extent of the neuraminic acid covering galactose or N-acetyl-galactose amino groups on glycoconjugates of the cell membrane. This correlation is substantiated by findings showing that mutants of the normally metastasizing B16 melanoma with little NeuAc and N-acetyl-glucose amine on their surfaces – demonstrated by wheat-germ agglutination (WGA) which reacts with NeuAc [11] – do not metastasize [417, 464]. The case is similar for other tumors [75–77]. There are, however, data which contradict this [295]. It is not known whether or not the increased binding of 'NeuAc-poor' tumor cells (mutants or caused by VCN treatment) to basal membrane collagen type IV and fibronectin is a decisive factor for lower metastasizing capacity [77].

The data available to date concerning a VCN-induced better exposure of naturally occurring antigens are contradictory. An increase in the antigenicity of nucleus-containing target cells after VCN treatment could not be demonstrated, neither using absorption tests [55, 67, 68, 147, 286, 330, 331, 336, 378, 379], nor with agglutination [216] or other binding studies [426], nor with the aid of lymphocyte cytotoxicity [269]. There have even been reports of complete destruction of an antigen associated with human lymphatic leukemia [233] or of a loss of immunogenicity in erythrocytes of different origin [25] after treatment with VCN. Moreover, there are divergent findings which point to an increase in the amount of antigen (recorded using antibodies) after VCN treatment: in human blood group antigens on leukemia cells [175], in HLA antigens [129], in IR (Ia) antigens [257], and in erythrocyte autoantigens [176]. When the transformation of autologous lymphocytes was used as a method of detection, similar findings could be recorded for tumor-associated antigens on AML-cells [296] and for cells of various mouse experimental tumors (3-LLAC; MC-squamous epithelial carcinoma [9]) and solid human tumors (bronchial carcinoma, melanoma, hypernephroma, adenocarcinoma of the breast and lung, stomach, colon, and rectum [6, 447, 448]). During transformation of autologous lymphocytes, however, it is difficult to distinguish an increase in antigenicity from a rise in immunogenicity, for instance due to membrane alterations, which can arise irrespective of changes in the tumor antigen.

In our own investigations, the influence of VCN was tested on various tumor-associated antigens, definable using monoclonal antibodies, on the cell membrane of culture cells, and on membrane-bound immunoglobulins of B-lymphocytes. After VCN treatment, there was neither a quantitative nor a qualitative change of the tumor-associated antigens, demonstrated with the use of monoclonal antibodies by immunofluorescence cytological tests, on the various tumor cells. After VCN treatment, the tumor cells reacted with monoclonal antibodies in the same way as before treatment (table III). Absorption tests were also unable to provide indications of a VCN-induced increase in the accessible TAA on the tumor cell membrane.

After VCN treatment of lymphocytes, a distinct increase in membrane immunoglobulin-positive lymphocytes could be demonstrated using the double antibody method [353, 369]. Tests with the relevant control antibodies showed an increase in unspecific binding varying in extent according to the donor. In certain cases, this corresponded with the rise in specific binding, so that for these donors, and in agreement with Lamelin et al. [204], no clear increase in membrane immunoglobulin-positive lymphocytes could be demonstrated. Both the increase in the number of lymphocytes stained with specific antibodies as well as the number of those non-specifically stained with control antibodies was reversible by cultivation for 20 h at 37°C (but not at 4°C) and could be inhibited by pretreating lymphocytes with an Fc-fragment from IgG (see table II). These findings point to an increase in the non-specific binding of antibodies used for immunofluorescence to the cell membrane after VCN cleavage of NeuAc, possibly via an increase in Fc-receptors.

An increase or decrease in antibody binding can also be caused by

Table III. Exposure of tumor-associated antigens on a small cell lung carcinoma cell (Oat-75) after treatment with VCN

MAB	VCN treatment[a]	Titer of[b] binding	Titer of binding after absorption[c] of MAB with different numbers of Oat-75 cells			
			10^4	10^5	10^6	5×10^6
227/18	−	32	32	32	4	
	+	32	32	32	4	
240/1	−	8				
	+	8				
252/104	−	32	32	16	8	4
	+	32	32	16	8	4
278/97	−	32				
	+	32				
278/105	−	32				
	+	32				
heteroserum	−	64				
	+	64				

[a] Oat-75 cells cultured in Terasaki plates; 100 mU VCN/well, 37°C; 30 min subsequently 3 washes

[b] Reciprocal value of titer endpoint. Detection by double antibody immunofluorescence technique using rabbit anti-mouse IgG, FITC-labeled.

[c] Absorption: cell sediment + 0.2 ml MAB (diluted 1:4 in PBS + 1% BSA); incubation 1 h; 4°C; subsequent centrifugation 300 g/10 min. Use of absorbed MAB for double antibody immunofluorescence technique.

MAB: Monoclonal antibodies according to Bosslet et al. [45]; starting dilution (1 µg/ml); Bosslet and Sedlacek (unpublished data)

receptor-independent alterations in cell membrane surface charge. In our experiments, this could be demonstrated after treatment and, thus, after incorporation of monosialo-ganglioside into the cell membrane of lymphocytes [370]. Depending on concentration and length of incubations, the incorporation of monosialo-ganglioside inhibited the specific binding of antibodies to membrane immunoglobulins of both untreated and VCN-treated lymphocytes. By means of a subsequent treatment with the B-fragment of choleragen, which possesses a high affinity for monosialo-ganglioside [410, 441], this inhibitory effect was donor-dependently completely reversible. Besides an increase in surface charge by monosialo-ganglioside as the cause of inhibition, membrane alteration or stabilization by monosialo-ganglioside is also, or additionally, possible. This stabilization could impede the movement of glycoproteins in the lipid bilayer of the membrane [50, 249] and, thus, their redistribution to form more easily recognizable 'spots', 'patches', and 'caps'.

The possibility of cell membrane-adherent VCN separating NeuAc from immunoglobulins may be seen as a further cause of non-specific adsorption of antibodies on VCN-treated lymphocytes. These immunoglobulins could become more cytophilic after separation of NeuAc [36].

In conclusion, it can be said that VCN treatment raises the unspecific binding of antibodies to the cell membrane, possibly due to exposure of Fc-receptors or by lowering the surface charge. This increase in non-specific binding may well simulate or mask an improved antigen accessibility, possibly due to NeuAc cleavage, and, thus, antigen detection. For various cell membrane antigens including selected TAA, however, no evidence for any VCN-induced demasking of already present cell membrane antigens could be found. This may depend on the type of antigens investigated. Thus, the degree of demasking of already present antigens could depend on how close together the antigen determinants and the NeuAc-rich zones influencing them are in the cell membrane [454]. The closer the contact between these two groups, the easier the access for antibodies or immune cells through NeuAc cleavage ought to be. In view of the local rotation and lateral movements of proteins and glycoproteins in the lipid double matrix of the membrane, in accordance with the lipid protein mosaic model [50, 249], the extent of demasking after VCN treatment should depend on the state of the cell membrane and cell type and, thus, be extremely susceptible to exogenous and endogenous factors.

Demasking of Neoantigens
(Thomsen-Friedenreich Antigen and Immune Response Antigens)

Separation of NeuAc can cause destruction of antigen determinants if NeuAc is an integral part of the epitope. This is the case, for instance, for an antigen associated with human lymphatic leukemia [233] as well as for the M- and N-blood group antigens [205], which are completely destroyed after treatment with VCN [399–401, 403]. These M and N antigens are glycoproteins which become Thomsen-Friedenreich (T) antigens after separation of NeuAc [104, 105, 421]. The immune-dominant structure of the T-antigen comprises disaccharide β-D-galactosido (1-3) α-N-acetyl-D-galactosamine [48, 183, 280]. This disaccharide is bound via its galactosamine to amino acids (serine, threonine) of the glycoprotein [125, 183, 279, 433, 434, 439].

Antibodies against the T-antigen can be normally found in a number of mouse strains [330, 331], in guinea pigs [331], and in normal humans [104, 105, 129, 168, 297, 298, 308, 315, 337, 434]. The origin of these autoantibodies is unclear. In humans the anamnesis excluded blood transfusions. Since free (that is, not masked by NeuAc) T-antigens should not appear in normal tissue [105, 404], it was speculated [103, 307–309, 337, 405–407] that anti-T antibodies originate in an immune reaction against carbohydrate antigens of micro-organisms or against autologous cells demasked by exogenous (from infection pathogens) or endogenous (lysosomal) neuraminidases. A further cause could be T-antigens on tumor cells developing due to an incomplete synthesis of oligosaccharide chains of membrane-bound glycoproteins [52, 99, 119, 134–136, 143, 178, 250], and even released into the surroundings from tumor cells [15].

Besides specific antibodies, cellular immune reactions could also be demonstrated in the course of the immune reaction against T-antigens (for instance in one-way mixed lymphocyte culture reaction (MLC)) [95, 100, 137, 169, 210, 214, 215].

It has been postulated that these immune reactions correlate with a successful tumor immunotherapy with VCN-treated tumor cells: i. e., the question has been raised as to whether or not the T-antigen arising after VCN treatment or due to incomplete oligosaccharide synthesis was a biologically and immunologically defined tumor-associated antigen with a decisive role in tumor immunotherapy − in other words, if an already present immune reaction against this antigen is a prerequisite for a therapeutically effective tumor therapy with VCN-treated cells. Thus Sanford and Codington [331] were only able to carry out a successful tumor immunotherapy with VCN in mice with antibodies against T-antigens. On the other hand, different tests [457] showed that the presence of these antibodies was not absolutely necessary, and an additional injection of these anti-T antibodies did not improve the immune therapeutic results [457]. Due to contradictory data, it is as yet unclear whether the titer of anti-T antibodies correlates in any way with tumor growth or is influenced by it [248, 307, 399, 400, 402, 404, 408, 409]. On the other hand, there appear to be indications, at least in mammary tumors in mice and various solid human tumors (bronchial carcinoma, gastrointestinal tumors, and especially in breast cancer [248, 404, 408]) that the T-antigen is tumor-associated and provokes a cellular immune response in the tumor carrier, which can be detected by testing delayed type hypersensitivity (DTH) with purified T-antigen [407–409].

Histological tests, especially on human mammary tumors, showed contradictory results. T-antigen could be found on malignant, but not on benign mammary tumors or normal gland tissue by means of sera containing antibodies against the T-antigen [15, 98, 157]. T-antigen could also be found in various transplantation tumors of mice and rats, and in the autochthonous mammary tumor in dogs [133, 362, 365] using peanut lectin (Arachis hypoglea) as well as sera containing anti-T antibodies. In those tumors, exposure of T-antigen could be enhanced by treatment of either cells or sections of the respective tumors with VCN [362].

On the other hand, after cleavage of NeuAc, the T-antigen could also be demonstrated on lymphocytes [48, 244, 246, 458], thrombocytes [121], human erythrocytes or those of the horse, sheep, cow [420, 439], or the dog [362]. Moreover, free T-antigens could be found on the surface of normal mammary gland cells, mainly on the epithelial cells of the ductuli and the lobules, and especially on active secreting cells, and furthermore, on milk fat droplet membranes [120, 245, 246], and this again may cast doubt on the tumor specificity of the T-antigen and the DTH reaction.

Other investigations with peanut lectin on human mammary tumors [186, 187] and on rat mammary tissue [247] showed that, in cases with invasive growth, tumor cells exposing free T-antigen move into the lumen of the supplying blood vessels. These tumor cells may possibly contribute to specify sensitization against the T-antigen. Peanut lectin reacts with T-antigens on various cells and organs in the same way as anti-T antibodies [174]. However, peanut lectin is not necessarily T-antigen specific since, on the one hand, it reacts with the disaccharide of the T-antigen (β-D-galactosido (1-3) α-N-acetyl-D-galactosamine) [40, 278, 433] and, consequently, recognizes a fragment of the T-antigen determinant [431], and, on the other hand, it can also react with other disaccharides (β-galacactose (1-3) α-galactosido structures and N-acetyllactosamine structures) under certain conditions [267, 400, 431].

Mean histocompatibility (MHC) class II immune response gene-coded antigens (Ia antigens) are a further group of membrane antigens which are partially susceptible to the enzymatic activity of VCN. Whereas the essential epitopes of the glycoproteins representing the Ia-antigens are situated in the protein fragment [64, 65, 346] thus offering no point of attack for VCN, there are additional Ia-antigens with the structure of glycolipids [151, 260–262] whose antigen determinants are determined by carbohydrates [260, 261] so that various monosaccharides and oligosaccharides are capable of inhibiting the binding of specific antibodies to glycolipid Ia-antigens [226]. Cleavage of NeuAc from these Ia-antigens representing carbohydrates destroys the serologically definable specificity of the initial Ia-antigen and causes both different and novel Ia-antigens to appear, which are not normally expressed [263]. Information from inhibition experiments with various sugars and enzymatic treatment with α-galactosidase points to the immune dominant structure of this demasked neoantigen as being a bound D-galactose in α-position. Of this it is known that it is a binding partner for NeuAc in many carbohydrates [19]. The role this demasking of neo-Ia-antigens could play is unknown; however, an increase in immunogenicity may be seriously considered.

In conclusion, VCN can expose new antigenic structures on the cell membrane by cleavage of terminal NeuAc from glycoproteins or glycolipids. At present it is unclear, however, whether or not those newly-formed epitopes represented by the linkage partner of the cleaved NeuAc can be regarded as TAA and are essential or effective in tumor immunotherapy.

The Appearance of Extraneous Antigens on Cells

In the course of our investigations into the working mechanism of VCN, the question arose whether VCN attaches to the membrane of cells. Using an antibody against VCN and immunofluorescence cytological tests in the double antibody technique, by determining the enzymatic activity (neuraminidase) of VCN-treated cells, and with the aid of radioactively labeled VCN (each technique was employed alone and in combination with others), it could be clearly demonstrated by us that despite excessive washing various amounts of VCN remain on the membrane of normal cells and tumor cells [213, 350]. This membrane binding was dependent on the enzymatic activity of VCN (heat inactivated VCN did not adhere) (tables IV and V). Moreover, cell membrane-bound VCN showed enzymatic activity.

These findings confirm the enzymological investigations of McQuiddy and Lilien [228], who showed that after VCN treatment and subsequent washings cells cleave considerably more NeuAc from a glycoprotein substrate than untreated cells. Furthermore, our findings are in agreement with the cell membrane binding studies of Nordling and Mayhew [251], who used FITC-labeled VCN. In contrast, Fidler et al. [97] using radioactively labeled VCN were unable to show any membrane binding.

A continuance of our investigations showed that the enzyme activity of cell membrane-bound VCN can be inhibited by specific antibodies.

Table IV. Detection of Vibrio cholerae neuraminidase on the cell membrane of VCN-treated cells

Cells	Treatments	Enzyme assay[a] (ext. 549 µm)	Immuno-fluorescence[b]
Erythrocytes (human)	—	0.14	—
	VCN	0.40	+
	VCN (100°C/10 min)	0.15	—
	VCN + AVCN	0.1	+
Peripher. lymphocytes (human)	—	0.05	—
	VCN	0.13	+
	VCN (100°C/10 min)	0.04	—
Molt 4 T	—	0.18	—
	VCN	0.2	+
RPMI 1788	—	0.14	—
	VCN	0.21	+

[a] 10^8 cells (sediment) + 25 mg orosomucoid in 1 ml Na acetate buffer incubation at 37°C for 30 min. Subsequently determination of free NeuAc by the method of Aminoff [79]

[b] Double antibody technique using rabbit anti-VCN antibody (AVCN)

VCN: 5×10^6 cells + 50 mU VCN in 1 ml PBS, 30 min/37°C, subsequently 4 washes According to Sedlacek and Seiler [350]

Table V. Detection of binding sites for radio-labeled VCN on the cell

Cells	Number of binding sites per cell	
	400 ng VCN (^{125}I)[a]	6.0 ng VCN (^{125}I)[a]
RPMI 1788	70,000	468
Peripheral lymphocytes	5,000/10,000	10/40
AML cells	3,000/25,000	5/10
Mammary adenocarcinoma cell line (dog)	10,000/50,000	30/40
Mammary adenocarcinoma (freshly prepared from canine tumor)	5,000/30,000	10/20

[a] Added to 1.5×10^7 cells and incubated at 37°C for 1 h
Specific radioactivity of VCN (^{125}I) = 4.7 Ci/g ≙ 0.2 ^{125}I atoms/molecule VCN
According to Lüben et al. [213, 371]

Surprisingly, however, the complex consisting of VCN and antibodies remained on the cell membrane despite enzyme inactivation so that the VCN binding to the cell membrane appears not to take place via the enzymatically active part of the VCN molecule. VCN binding to the cell membrane was greatly reduced by pretreating the cells with β-galactosidase or trypsin. Moreover, it was competitively inhibited by the addition of lactose to the incubation mixture of cells and VCN (see table VI), to a lesser extent by galactose, sucrose, or glucose [342]. Thus, we think it is probable that the membrane structure to which VCN binds is glycoprotein containing a terminal galactosid or β-D-galactosido (1-4)-β-D-glucosid. Since galactose is the glycosidic binding partner of NeuAc and is expressed after separation of NeuAc, the enzymatic activity of VCN could lead to an increase in membrane binding points for VCN. This assumption would offer an explanation for our findings that microorganisms, such as BCG and E. coli, from the surface of which no NeuAc can be cleaved, cannot be shown to bind any amounts of VCN to their surface [342].

The observation that VCN binds to the cell membrane and, when bound, is enzymatically active, was of interest for a number of reasons. For one, VCN is an exogenous antigen on the cell membrane which can alter both the antigenicity and the immunogenicity of the cell considerably. For another, there is the possibility of membrane-bound VCN

Table VI. Influence of pretreatment of SRBC on subsequent cell membrane binding of VCN

Pretreatment	Treatment with VCN in mixture with	Release of NeuAc from orosomucoid by cells (%)[b]
	VCN	100
β-galactosidase		
Exp. 1 25 U/ 30 min	VCN	35
25 U/ 60 min	VCN	45
25 U/120 min	VCN	40
25 U/180 min	VCN	9
Exp. 2 1 U/ 60 min	VCN	31
5 U/ 60 min	VCN	5
25 U/ 60 min	VCN	5
Fucosidase		
0.50 U/60 min	VCN	100
Mannosidase		
2.50 U/60 min	VCN	112
Trypsin		
0.25 U/60 min	VCN	5
–	VCN + lactose[a]	29
–	VCN + galactose	50
–	VCN + sucrose	50
–	VCN + glucose	64
–	VCN + xylose	86
–	VCN	100

Pretreatment: 10^8 SRBC (sediment) + 0.2 ml enzyme in PBS; incubation at 37°C, subsequently 3 washes in PBS
Treatment with VCN: 10^8 SRBC in 0.1 ml PBS + 50 IU VCN (0.1 ml); incubation at 37°C for 30 min, subsequently 3 washes
[a] Concentration of sugars 10 µMol
[b] 10^8 SRBC + 25 mg orosomucoid/ml; incubation 30 min/37°C, subsequently assay of free NeuAc according to Aminoff [13]
According to Schneider et al. [342]

coming into contact with neighbouring cells due to its enzymatic activity and possibly developing adjuvant activity. Reasons for this assumption were given by the investigations carried out by Petitou et al. [268].

These investigations demonstrated that cell membrane-bound VCN is capable of separating NeuAc from partner cells, such as lymphocytes, which are brought into close contact with VCN-bearing cells.

We investigated, in particular, the role of VCN as an exogenous antigen on cell membranes [168, 171]. Cytotoxic antibodies against VCN-treated cells could be found in the serum of normal persons and patients (multiple sclerosis, tumor illnesses) relatively frequently (30 to 48% of all cases), but not in blood from the umbilical cord of newborn babies. These antibodies could not only be selectively absorbed by VCN-treated cells (lymphocytes, erythrocytes, tumor cells) but also by VCN covalently bound to carriers (polyhydroxymethyl or Sepharose 4B). They could be eluted from the carrier and inhibited with lactose. In addition, these eluted antibodies inhibited to varying extents the enzymatic activity of VCN and influenza virus neuraminidase, but not that of the Clostridium perfringens neuraminidase [171]. Since, additionally, in a number of test persons both active and heat-inactivated VCN spontaneously stimulated lymphocytes to transformation, it may be assumed that a considerable number of normal people and patients have an acquired (direct) or cross-reacting immunity against VCN, which can play an as yet unknown role in the question of the immunogenicity of VCN-treated cells and especially tumor cells, as VCN sticks to the membrane of VCN-treated cells and may function there as an extraneous antigen.

The Influence on the Immune System

Influence on Cell Immunogenicity

Numerous investigations have shown that an increase in immunogenicity of cells can be caused by VCN treatment. All these investigations are based on the original observations of Lindemann and Klein [211] that an injection of VCN-treated Ehrlich-ascites tumor cells can protect mice from the growth of a subsequently injected large number of the same tumor cells. At the same time as Lindemann and Klein [211], Sanford [329] and Currie [66] observed a decrease in tumor-cell transplantability after VCN treatment. Moreover, these study groups found a specific protection against subsequent tumor transplants in mice injected with VCN-treated tumor cells, inactivated in their proliferative capacity by treatment with mitomycin or by irradiation.

Control investigations showed [329] that the observed phenomena had not been caused by non-specific activation of the reticuloendothelial system. Application of complete Freund's adjuvant (CFA) alone proved to be ineffective, and the survival periods for allogeneic skin transplants were not influenced by administration of VCN-treated syngeneic tumor cells or by VCN. These initial findings in selected tumors were able to be confirmed by a number of study groups in various tumor systems (see table VII). In these investigations, immunity was found to be tumor-specific [373, 380, 381, 382]. However, cross-reactions could be observed, for instance, in allogenous E_2G-leukemia cells, which, like AKR-leukemia cells, are induced by Gross virus, and which were equally effective in protection tests as syngeneic AKR-leukemia cells [29, 30, 154]. This is possibly a result of a common virus-induced membrane antigen.

The increased immunogenicity of tumor cells after VCN treatment was demonstrated by a decrease in tumor cell transplantability [28, 68, 69, 92, 292, 293, 380, 383], by resistance to a second injection of (living) tumor cells [7, 29, 30, 91, 92, 166, 292, 380, 381] (for further information,

Table VII. Increase of tumor cell immunogenicity after VCN treatment (demonstration by transplantation tests)

Ehrlich ascites	mouse	Lindemann and Klein [211]; Sur and Roy [413]
TA$_3$ ascites adenocarcinoma	mouse	Sanford [329]
Landschütz ascites	mouse	Currie [66]; Brazil and McLaughlin [49]; Smyth et al. [394]
L 1210 leukemia	mouse	Bagshawe and Currie [22]; Bekesi et al. [28]; Brandt et al. [46]
AKR leukemia	mouse	Bekesi and Holland [30]
MC-42, -43, -10	mouse	Simmons and Rios [380, 387]; Faraci et al. [92]
Mast cell tumor	mouse	Sedlacek and Seiler [351]
Ependymoblastoma (GL-26)	mouse	Albright et al. [7]
DMBDN fibrosarcoma	mouse	Ray and Sundaram [290]; Ray et al. [292, 293]
B 16 melanoma	mouse	Jamieson [166]; Brinkerhoff and Lubin [51]
Ni$_3$S$_2$-fibrosarcoma	rat	Abandowitz [1]
E$_2$G	mouse	Bekesi and Holland [30]

see table VII), by an increase in specific antibodies [67, 68], in specific lymphocytes cytotoxicity [7, 91] or macrophage cytotoxicity [10], by transformation of autologous lymphocytes [6, 9, 296, 446–448], by inhibition of immunosuppression [67, 68, 293, 373, 383], and by transfer of the induced immune protection, with the aid of serum or spleen cells, to syngeneic, non-immunized mice [28–30, 69, 282, 302, 304, 305, 373, 374, 383, 386, 387].

The immunogenicity of various normal cells could also be increased by VCN treatment: that of the stimulating lymphocytes in one-way MLC [97, 214, 215] and of SRBC in plaque-forming cytolysis (PFC) assay [338–340].

Other investigations showed that immunization of mice with allogenous VCN-treated embryonal cells, lymphocytes, spleen, or bone marrow cells, in contrast to untreated cells, led to a more rapid rejection of skin grafts, whereby the skin grafts were syngeneic with the cells used for immunization [161, 382]. Furthermore, the 'graft-versus-host' disease could be reduced in animals treated sublethally with cyclophosphamide by prior treatment with VCN-treated spleen cells syngeneic with the 'graft' [161]. This could point to an increase in the immunogenicity of VCN-treated cells.

However, there are investigations which produced quite different results, in which − by means of immunization or in lymphocytotoxicity − no increase in immunogenicity could be shown after VCN treatment of, for instance, SRBC and erythrocytes of other species [25] or tumor cells (sarcoma and mastocytoma [426]).

An increase in immunogenicity after treatment with VCN was clearly independent of the viability or tumorigenicity of the treated tumor cells. Thus, radiation or treatment with cytostatic drugs impaired the tumorigenicity of VCN-treated cells, but not their increased immunogenicity. On the other hand, VCN-treated tumor cells proliferated as well as untreated tumor cells in immune suppressed syngeneic animals [293, 387, 388]. Tumor cell membrane preparations were also found to be immunogenic after VCN treatment as revealed by specific protection experiments [46, 272, 375].

In a comparison of neuraminidase of different origins, VCN was found to be superior to Clostridium perfringens neuraminidase (CPN) or influenza virus neuraminidase (IVN) with regard to cleavage of NeuAc from the cell membrane or an increase in tumor cell immunogenicity [30, 67, 68]. With CPN, an increase in immunogenicity could be achieved, which was as good as that after VCN treatment, only after a long incubation period [67]. IVN, however, cleaves only the 2−3 and 2−8, but not 2−6 ketosidic linkage of NeuAc [86] and, therefore, appears to be (nearly) ineffective to increase immunogenicity of cells [30, 288]:

The increase in cell immunogenicity was dependent on the enzymatic activity of the VCN used to treat the cell: heat inactivation [30, 381] or inhibition of the enzyme by addition of NeuAc [380−382], by pH-alteration [30], or by cooling to 4°C [68] reduced the ability of the enzyme to increase cell immunogenicity.

During immunization experiments, a dependence could surprisingly be found between the amount of VCN used to treat the tumor cells and the immunogenicity of these tumor cells. If tumor cells, inactivated by irradiation or treatment with mitomycin, were treated with VCN concentrations two or three times higher than the optimal amounts for immunization (50−200 mU VCN/2.5×10^7 L1210-cells/ml or 5−100 mU VCN/10^6 sarcoma cells/ml), the immunogenicity of L1210 leukemia cells [30, 373] or of methylcholanthrene-induced fibrosarcoma cells [381, 383] was reduced. This fall in immunogenicity after incubation with larger amounts of VCN can neither be explained enzymologically [357] nor can it be attributed to a cytotoxic effect of VCN [159, 299, 315, 329]. Nevertheless, there is still the possibility that the dependence of tumor cell immunogenicity on the VCN treatment dose is caused by the different amounts of VCN which remain on the cell membrane despite washing and then act as an adjuvant. It is astonishing that similarities can be seen between the dose effect curve of the VCN used for cell treatment and the resulting immunogenicity of the VCN-treated cell on the one hand and the dose effect curve in our adjuvant investigations with VCN on the other.

Adjuvant Activity of VCN

In order to clarify the question of whether or not cell membrane-bound VCN can work dose-dependently as an adjuvant for cells, we carried out immunization tests with SRBC as a cellular model antigen, mainly using a method developed by Mackaness et al. [219, 220]. When selecting SRBC, it was taken into consideration that after VCN treatment, these cells have Thomsen-Friedenreich antigens on their surfaces [439]. Different numbers of SRBC, either treated with VCN or mixed with different amounts of VCN, were injected intravenously or subcutaneously into mice and the strength of the immunization was evaluated by determining the DTH reaction on a subsequent intraplantar injection of a constant number of SRBC and by titration of hemolytical antibodies.

The admixture of VCN, but not VCN treatment, clearly enhanced the specific cellular immune reaction (DTH) against SRBC [354–356]. Whereas no VCN-induced titer increase of hemolytic antibodies could be shown, the number of cells in the spleen specifically forming antibodies against SRBC (plaque forming cells) increased [191]. This increase, mainly of the DTH reaction, but also − even if only to a limited extent − of the antibody response to SRBC, only took place when VCN was injected mixed with the antigen [191, 356]. No adjuvant or immunostimulating effect was achieved when VCN was injected separately from SRBC, either simultaneously or at different points in time. The adjuvant activity of VCN was dependent on the enzymatic activity of the enzyme, since heat inactivation of VCN or addition of NeuAc, well known as a VCN inhibitor [232, 442], destroyed or inhibited its adjuvant activity [191]. Moreover, this activity was dependent on an optimal amount (mostly 5–100 mU VCN) of the enzyme (smaller or larger amounts had less effect), on the number of cells used for first immunization, and the method of application. After i. v. injection, VCN increased the DTH reaction in mixtures with predominantly low or moderate cell numbers ($10^3 - 10^6$ SRBC), whereas after s. c. injection, the DTH response was enhanced after application of higher cell numbers ($10^6 - 10^9$). Repeated administration of VCN, for instance, added to SRBC during the second immunization, tended rather to reduce the DTH reaction to the second injection of SRBC [356].

In order to clarify the question of whether or not the adjuvant activity of VCN is independent of the cleavage of cellular NeuAc on the immunogen, i. e., of a demasking of membrane-bound antigens, adjuvant tests were carried out by us with proven neuraminic acid-free immunogens (E. coli, BCG). Similarly, as in the case of SRBC, VCN increased in a mixture with E. coli or BCG the humoral (E. coli) or cellular (BCG) specific immune reaction (see table VIII) whereby this adjuvant effectiveness also was dependent on the enzymatic activity and the VCN dose [191, 372].

In this context, it was surprising to see that several concentrations ($10^6 - 10^7$) of living BCG germs, administered 12–17 days before application of the test antigens SRBC or VCN (a period of time after which BCG is known to be most immunostimulating) [220], nonspecifically increased the immune reaction not only against SRBC but also against VCN, but, when administered in a mixture with VCN, did not possess any additional adjuvant activity. In contrast, when administered in a mixture with living BCG germs, VCN increased the immune reaction against BCG [372]. VCN itself must be considered as an extremely weak immunogen [310], whereas BCG, in accordance with the Mantoux-Test in BCG-vaccinated children, has to be regarded as a relatively strong immunogen.

Thus, BCG is not capable of adjuvating a weak immunogen like VCN, whereas VCN displays such a strong adjuvant activity that the immune reaction against a strong immunogen such as BCG is enhanced even more.

As is known, treatment of animals with living BCG enhances the

Table VIII. Adjuvans activity of VCN for the in-vivo immune response against different antigens

Antigen	VCN optimal dose range (mU)	Species	Maximal increase (% to control)	Assay system
SRBC[a]	5	mouse	125	DTH (Mackaness et al. [220])
SRBC[a]	0.5–50	mouse	390	PFC (Jerne et al. [167])
E coli[b]	0.5–50	mouse	700	bactericidal test (Rowley [320]; Heddle et al. [148])
BCG[c]	5	mouse	55	DTH
BCG[d]	0.5–50	guinea pig	45	DTH
			130	Ly-Transf. (Dubois et al. [88])
S. typhi murium[b]	5	mouse	690	passive hemagglutination assay (Crumption et al. [62])
Rubella virus[b]	5	mouse	460	Passive hemagglutination inhibition (Stewart et al. [411])

[a] Knop et al. [191]
[b] Sedlacek and Seiler [356]
[c] Seiler and Sedlacek [372]
[d] Johannsen et al. [170]

immune reaction non-antigen-specifically. Injection of VCN mixed with SRBC in animals pretreated with BCG led to no further increase in the DTH reaction or the antibody reaction (see table IX). Thus, VCN is ineffective as an adjuvant in animals nonspecifically immunostimulated with BCG [354, 356].

All in all, it can be concluded that VCN works as an adjuvant when it is injected together with an immunogen. The adjuvant effectiveness of VCN is, on the one hand, independent of NeuAc cleavage from the immunogen and, on the other hand, clearly dependent on the enzymatic activity of the enzyme and on the presence of the antigen at the place of injection. Thus, it can be assumed that VCN works as an adjuvant due to its enzymatic activity on cells in the immune system, possibly influencing the contact with or processing of the immunogen by these cells.

Table IX. Adjuvant activity of VCN for the delayed-type hypersensitivity reaction against SRBC

Pretreatment (day −12)	VCN in mixture with SRBC (day 0)	Application[a]	DTH response (footpad swelling in % of original value)[b]						
			10^3	10^4	10^5	10^6	10^7	10^8	10^9
—	—	i.v.	25 ± 1	25 ± 3	38 ± 5	47 ± 3	40 ± 4	29 ± 3	21 ± 3
—	5 mU	i.v.	30 ± 5	36 ± 3	60 ± 4	50 ± 3	44 ± 4	39 ± 4	31 ± 3
—	5 mU (100 °C; 15 min)	i.v.	24 ± 3	29 ± 4	48 ± 7	44 ± 3	40 ± 2	38 ± 4	29 ± 1
—	—	s.c.	22 ± 3	17 ± 2	27 ± 5	26 ± 3	34 ± 3	45 ± 4	46 ± 7
—	5 mU	s.c.	21 ± 1	21 ± 5	29 ± 3	36 ± 3	44 ± 5	57 ± 5	60 ± 5
—	5 mU (100 °C; 15 min)	s.c.	21 ± 3	24 ± 2	28 ± 7	34 ± 3	41 ± 2	40 ± 3	51 ± 4
—	—	i.v.	34 ± 6	31 ± 6	39 ± 3	35 ± 6	45 ± 6	37 ± 4	37 ± 4
BCG	—	i.v.	n.d.	49 ± 4	54 ± 4	74 ± 8	n.d.	55 ± 8	n.d.
BCG	5 mU	i.v.	30 ± 6	33 ± 7	33 ± 9	44 ± 8	45 ± 7	45 ± 8	52 ± 11

BCG: 10^7 living BCG (0.01 ml) on day −12

[a] Injection of 10^3–10^9 SRBC (0.15 ml) ± 5 mU VCN on day 0

[b] s.c. injection of 2×10^8 SRBC (0.05 ml) into the footpad on day 5; measurement of footpad before and 24 h after injection. m ± s of 10 animals per group

According to Sedlacek and Seiler [356]

Influence on Granulocytes, Macrophages, and Lymphocytes

To answer the question as to which particular cells are influenced by VCN, in-vitro activation tests were carried out with various leukocytes.

Macrophage or granulocyte activation was measured by the chemoluminescence reaction which is known to be induced by the development of superoxides in the course of a rise in oxygen consumption and glucose oxidation via the hexose-monophosphate pathway [20, 184, 318, 333]. Moreover, activation of macrophages and granulocytes was evaluated by quantifying exocytosis of lysosomal enzymes [377], such as N-acetylglucose-aminidase or β-galactosidase, known to be linked to phagocytosis [160, 164, 325, 418, 436]. However, exocytosis is not always accompanied by phagocytosis [123, 146, 149, 150].

The results are contradictory. VCN did not activate macrophages or granulocytes in vitro or in vivo, when activation was recorded [111] by means of chemoluminescence and exocytosis of lysosomal enzymes (see tables X, XI). In contrast, Suzuki et al. [414] found an increase, Tsan et al. [423, 424] an inhibition of chemoluminescence after treatment with Clostridium perfringens neuraminidase and (similar to Noseworthy et al. [253]) an inhibition [414] of the phagocytosis rate. These differences in results are possibly due to different degrees of purity of the various neuraminidase preparations used. In this respect, contamination of lipopolysaccharides seems to be of particular significance.

On the other hand, measured by oxygen consumption [414], the release of peroxides induced by purified Clostridium perfringens neuraminidase was not accompanied by an increased cell metabolism, so that granulocyte activation is improbable and membrane damage is more likely to be the cause of peroxide release. This is also indicated by the fact that the phagocytosis rate was not influenced despite increased chemoluminescence [414].

Our experiments, in which macrophages could be stimulated with VCN either to increased exocytosis or to increased peroxide formation, may contradict results of other tests carried out by us which demonstrate an increased intraperitoneal 'clearance rate' of E. coli germs in mice after i.p. injection of VCN [190]. Inhibition tests with intraperitoneally injected dye or with orosomucoid, cytological investigations of the peritoneal exudate, and/or control tests with heat-inactivated VCN

Table X. Activation by Vibrio cholerae neuraminidase of macrophages and granulocytes

VCN (mU)	Mouse peritoneal macrophages							
	RLU	in vitro N-acetyl-glycosaminidase (U/ml)		RLU	ex vivo β-galactosidase (U/ml)			
		total	supernatant		total	supernatant		
—	350 ± 24	3.6 ± 0.3	0.9 ± 0.2	1688 ± 177	11.3 ± 0.5	1.2 ± 0.3		
50	376 ± 22	3.0 ± 0.3	1.1 ± 0.1	1700 ± 184	12.2 ± 1.0	1.8 ± 0.6		
25	287 ± 52	3.0 ± 0.4	0.8 ± 0.3	1910 ± 153	10.7 ± 0.6	1.3 ± 0.2		
10	368 ± 33	4.1 ± 0.4	1.2 ± 0.2	1840 ± 71	11.2 ± 0.6	1.3 ± 0.1		
5	294 ± 74	3.5 ± 0.5	0.8 ± 0.1	1635 ± 107	11.4 ± 0.7	1.3 ± 0.2		

RLU: Relative light units ($\times 10^3$) emitted by 10^6 macrophages or 10^5 granulocytes (measured in Biolumate); VCN added to cells immediately before measurement (in vitro)

ex vivo: i.p. injection of VCN in 0.5 ml PBS; 72 h later harvesting of peritoneal macrophages and measurement of RLU or exocytosis

Determination of N-acetyl-glycosaminidase and β-galactosidase from 3×10^6 cells after 24 h culture (m ± s of 4 mice)

Methods have been performed as described in [343], Schorlemmer and Sedlacek, unpublished data

Table XI. In-vitro activation by Vibrio cholerae neuraminidase of human granulocytes and cultured monocytes

		VCN (mU)	Monocytes	Granulocytes
Donor I		—	2,105 ± 107	607 ± 12
		50	2,100 ± 141	870 ± 198
		5	2,405 ± 545	702 ± 99
		0.5	1,910 ± 226	852 ± 101
	IC		29,150 ± 778	25,850 ± 475
Donor II		—	496 ± 63	
		50	522 ± 29	
	SRBC		122 ± 23	
	SRBC +	50	109 ± 40	
	IC		6,420 ± 107	

SRBC: 1×10^8 SRBC per test
RLU: Relative light units ($\times 10^3$) (measured in Biolumate), emitted by 10^6 monocytes or 10^5 granulocytes
VCN added to cells immediately before measurement
IC: Tetanus toxoid – antitetanus toxoid antibody – complexes at equivalence
Methods have been performed as described in [343]; Schorlemmer and Sedlacek, unpublished data

pointed to macrophages as the responsible cells and to the necessity of VCN enzymatic activity.

The enhanced clearance, however, can be interpreted in such a way that VCN treatment of peritoneal macrophages reduces the macrophage negative surface charge, increases the adsorption of E. coli germs to the cell membrane [190], and increases phagocytosis of those germs. As E. coli germs neither absorb VCN [191] nor do they have VCN-cleavable NeuAc [79], any effect of VCN on E. coli can be excluded. This assumption would be in agreement with the findings of Weiss et al. [451, 454], who saw an increased accumulation and phagocytosis of plastic particles after VCN treatment of human monocytes, and may explain similar findings on the phagocytosis of SRBC [207, 338, 339] or B. subtilis [8].

VCN treatment of lymphocytes raised their transformation rate induced by (table XII) mitogens (Concanavalin A; pokeweed mitogen) [37, 117, 138, 141, 170, 254, 327] or by antigens (PPD, rubella, tetanus toxoid, influenza) [37, 138, 140, 141, 170, 266], and increased cellular and humoral cytotoxicity against several target cells [94, 111, 177, 454]. An

Table XII. Increase by VCN of antigen or mitogen-induced lymphocyte transformation in vitro

VCN (mU)	0	0.5	5
PPD	6.5 ± 2.1	8.9 ± 0.9	13.0 ± 1.9
Rubella	3.0 ± 1.0	5.2 ± 1.4	11.3 ± 1.2
Tetanus	4.8 ± 0.5	6.4 ± 0.2	8.1 ± 1.4
Influenza	3.9 ± 0.9	5.4 ± 1.6	7.8 ± 2.4
—	0.3 ± 0.08	0.4 ± 0.1	0.4 ± 0.01
LPS (10 µg)	0.5 ± 0.2	0.6 ± 0.1	0.6 ± 0.1
ConA (8 µg)	14.9 ± 2.3	20.0 ± 1.2	22.7 ± 1,8
PWM	4.8 ± 0.5	6.8 ± 0.4	11.4 ± 0.6

3×10^5 human peripheral blood lymphocytes were incubated with 0.1 ml of the various antigen preparations or with different amounts (as indicated) of mitogen preparations ± increasing amounts of VCN in a total volume of 1 ml RPMI 1640 for 5 days. On day 5, ^{14}C-thymidin was added and 24 h later incorporation of radio-labeled thymidin into the lymphocytes was measured according to DuBois et al. [88]
According to Johannsen et al. [170]

increase in the antigen-specific reactions was only detectable in already sensitized lymphocytes, which would rule out an effect due to a general alteration of surface charge [454]. On the other hand, lymphocytes which increasingly bind to tumor cells [111] can increase their nonspecific NK-cell activity after VCN treatment.

The 'adjuvant' effect of VCN in lymphocyte transformation has to be seen as quite separate from an additionally appearing and possibly cross-reactive sensitization against VCN itself in various lymphocyte donors [170, 172].

The 'adjuvant' effect of VCN in lymphocyte transformation could be shown by us particularly clearly in lymphocytes from donors with low transformation rates against the various antigens, and less clearly in the case of strong reactants. Using the example of lymphocyte transformation by the antigen PPD (purified protein derivative of mycobacterium), it was possible to further show that the 'adjuvant' effect of VCN could only be demonstrated when VCN was admixed to the lymphocyte culture simultaneously with or shortly after (24 h) administration of the antigen [170].

If the antigens, required for lymphocyte transformation, were incubated with carrier-bound (covalent binding to polyhydroxymethylene)

VCN, subsequently separated and added to the lymphocytes, we could find no increase in lymphocyte transformation [170] which would indicate a direct effect of VCN on lymphocytes.

Findings in mixed lymphocyte culture reaction (MLC) are somewhat in contrast to the above, in that VCN treatment of the cell functioning as an immunogen, but not of the transforming cell, leads to an enhanced stimulation [97, 214, 215], whereas Han [137] was able to find increased stimulation in the MLC after VCN treatment of both cells.

It is evident from our experiments that VCN assisted the immune response against cellular immunogens and against bacteria [170, 191, 356]. In comparative studies on guinea pigs we could show [170] that the increase in the DTH reaction against PPD in animals immunized with VCN and BCG was accompanied by an increased transformation rate of spleen cells after exposure with PPD. These results are basically similar to those of Han [139, 141] and Pauly et al. [266] who were able to observe a VCN-induced increase in immune response (secondary reaction) against various antigens in specifically sensitized persons in vivo (DTH reaction) and in vitro (lymphocyte transformation).

In both types of reaction, T-lymphocytes are decisively involved, and both test systems reflect the adjuvant activity of VCN for the specific immune reaction.

The mechanism underlying this adjuvant effect of VCN on lymphocytes is unclear. However, the studies carried out by Knop [188, 189], in view of our results, point to an increased lymphocyte sensitivity induced by VCN treatment for mediators produced by macrophages, for instance, the lymphocyte-activating factor (LAF). As is known, macrophages present in lymphocyte preparations are necessary for lymphocyte transformation as a reaction, for instance, to Concavalin A (ConA) [17, 235]. Removal of these macrophages leads to a drastic reduction in lymphocyte transformation, which can be recompensed by admixture of supernatants (LAF) of activated macrophages [188, 189]. In this model, an increase in lymphocyte transformation could be achieved by admixture of VCN. Inhibition tests with N-acetylneuraminyl-lactose showed that VCN does not effect the LAF but the susceptible T-cell [188, 189].

VCN did not only sensitize T-cells for the activating, but also for the inhibiting factors of macrophages [188, 189], which may explain the findings of Pauly et al. [266] and Adler et al. [5] who found a VCN inhibition of the lymphocyte transformation induced by ConA or phythemagglutinin (PHA). Since the VCN-induced sensitization for inhibiting factors could not be blocked by indomethacine, an influence of VCN on the effect of prostacyclins may be ruled out [188, 189].

Various mechanisms may be discussed for the VCN-induced sensitization of T-cells for mediators: less repulsion between mediator and receptor by a reduction of surface charge in T-cells, moreover, a demasking or better accessibility of the receptors.

VCN not only influences the receptor on T-cells for mediators. There are indications that after VCN treatment bone marrow stem cells become more susceptible to erythropoietin [273, 321] and that VCN treatment reduces the insulin-dependent glucose transport in fat cells, possibly due to an increase in signal transmittance after insulin binding to its receptor [63]. Our findings that number and function of Fc-receptors on lymphocytes is drastically increased after VCN treatment [369] may point in the same direction.

In summary, it can be concluded that VCN, by affecting lymphocytes, has an adjuvant activity for the immune response against different antigens. This adjuvant effect depends on the enzymatic activity of VCN and is dose dependent (bell-shaped dose response curve). VCN showed no effect as an adjuvant [356] in animals nonspecifically stimulated with bacillus Calmette-Guérin (BCG). The reason for this is unclear, but it can be assumed that BCG pretreatment had already in a non-antigen-specific way increased those mechanisms to a maximum, which are also stimulated by VCN as adjuvant for the specific immune reaction. Despite its strong adjuvant activity (in comparison, for instance, with BCG) [361], VCN was not able, when injected in a mixture with myelin as an immunogen alone or in additional combination with incomplete Freund's adjuvants (IFA) or CFA, to cause or to aggravate the experimental autoallergic encephalomyelitis in rats [258, 264, 265], neither could VCN do this in animals which had been unspecifically immunostimulated by pretreatment with living BCG germs. This lack of effect of VCN may be regarded as an indication of its immunotoxic harmlessness.

Adjuvant Effect for Tumor Cells

The immunogenicity of different tumors can be increased by treatment of those tumor cells with VCN (table VII). For instance, in our own investigations [351] on mastocytoma, 70% of all animals, pretreated with 1.5×10^2 VCN-treated tumor cells inactivated in their growth potential by incubation with mitomycin, survived a subsequent transplantation of 0.5×10^2 mastocytoma cells, whereas, in contrast, all the animals in the non-pretreated control group or 90% of the animals pretreated with 1.5×10^2 untreated inactivated tumor cells died. Mastocytoma grows locally and is an immunogenic tumor [329]. Considering the adjuvant effect of VCN, however, the question was whether or not the immunogenicity even of weakly immunogenic tumor cells can be enhanced not only by VCN treatment but also and possibly to a much higher degree by adding VCN as an adjuvant to tumor cells. To answer this question, we carried out investigations [360, 362] on the weakly immunogenic and metastasizing 3 LL AC [227, 440]. Primary tumors were transplanted into mice and their growth and progression times recorded. At a point in time when, in accordance with preceding control examinations, a micrometastasation had taken place, the primary tumor was removed and the survival time of the animal recorded.

Pretreatment of the animals with different numbers of either mitomycin-inactivated tumor cells or of inactivated tumor cells, which had been either treated with VCN or mixed with various amounts of VCN, had no influence on primary tumor growth.

Moreover, the metastasis-induced mortality was not reduced by pretreatment with different numbers of inactivated tumor cells; pretreatment with different amounts of VCN-treated inactivated tumor cells also had no or only a marginal effect. In contrast, after application of a mixture of 1×10^6 inactivated tumor cells with 100 mU VCN, metastasis-caused mortality was clearly reduced, but in one experiment only. Smaller amounts of VCN (50 mU and 10 mU) proved ineffective. In a repeated experiment, a prophylactic protective effect on metastasis-induced mortality could only be seen when a 5-fold amount (5×10^6) of inactivated tumor cells mixed with 50 mU VCN was injected twice, with 14 days between injections. In similar experiments with heat-inactivated VCN, no influence on mortality could be found [360, 362].

Thus, it might be concluded from these experiments that immunization with VCN-treated inactivated tumor cells in a weakly immunogenic metastasizing tumor model is inferior to that with a mixture of tumor cells and VCN. The effect of this mixture, however, is dependent on the amount of enzymatically active admixed VCN, possibly caused by its adjuvant function. More importantly, difficulties may arise when mixtures are administered, in reproducing a protective effect once observed with a certain number of tumor cells and a certain amount of VCN. No explanation for this lack of reproducibility can be given. It is possibly a reflection of quantitative differences, which we cannot register, in the properties of the tumor cell preparations used for immunization or of different constitutions and conditions in the animal populations used for experiments.

Experimental Tumor Therapy Tests

Administration of VCN-Treated Inactivated Tumor Cells

VCN treatment increases the immunogenicity of tumor cells inactivated in their proliferative capacity by treatment with cytostatics or by irradiation. Similar results could be found for VCN-treated cell membrane preparations from tumor cells [46, 272]. Immunogenicity has mainly been elaborated by specific protection of mice against a subsequent challenge with living tumor cells. Based on these protection experiments, a therapy for solid tumors (first by Simmons et al. [380, 381]) and for experimental leukemia (first by Bekesi et al. [28]) was attempted by applying VCN-treated inactivated tumor cells with various rates of success. In a number of investigations by various authors, inhibition of tumor growth, total tumor regression (within a period of 30 days), or a lengthening of the survival time could be observed (see table XIII). By analyzing all those experiments [357, 364] the following conditions were found to be prerequisites for successful therapy:

– The tumor cells used for therapy have to be identical with the tumor cells to be treated or, at least, have to have the same antigens since non-identical tumor cells were ineffective [28, 31, 380, 381].

– The VCN used in therapy has to be enzymatically active. Injections of tumor cells which had been treated with heat-inactivated VCN or VCN inhibited by addition of high doses of NeuAc or neuraminyl-lactose did not impair tumor development [30, 386].

– The tumor which is to be immunotherapeutically influenced has to be as small as possible. In the tumor models investigated, tumors with a diameter of more than 1 cm [304] or 0.5 cm [274] did not shrink despite immunotherapy. Methods of reducing tumor volume, such as surgical removal of a part of the tumor mass [304, 457], were therapeutically more effective in combination with immunotherapy than any of the methods of treatment alone. However, in the case of strongly metastasizing and infiltratively, fast-growing tumors (B16 melanoma) even the combination could not produce any additional effect, possibly due to an ineffective surgical treatment [304, 305].

– The tumor-bearing animal has to be immunocompetent. Reduction of the tumor mass by surgery [304, 305], chemotherapy [29, 69], or radiation [395] has to be selected and dosed in such a way that the immune system is still able to react specifically and effectively to stimulation by an immunogen [29, 69, 380, 382, 386, 387].

Treatment with an excessive dose of cytosine arabinoside and procarbazine, for instance, prevented the immunotherapeutic effect of

Table XIII. (a) Effective, (b) ineffective tumor therapy with VCN-treated tumor cells

Type of tumor	Histology	Type of animal	References
(a)			
Chemically induced non-metastasizing	fibrosarcoma (various types)	mouse	Simmons et al. [380, 381] Rios and Simmons [282, 303–304]; Ray et al. [293]; Ray and Seshadri [294]; Wilson et al. [457]; Song and Levitt [395]
	Yoshida sarcoma	rat	Ray and Seshadri [294]
	squamous epithelial carcinoma	mouse	Alley and Snodgrass [9]
Transplantable, metastasizing unknown origin	B 16 melanoma	mouse	Rios and Simmons [303]
	Lewis lung adenocarcinoma	mouse	Alley and Snodgrass [9]
	Dunn osteosarcoma	mouse	Miller et al. [234]
	melanoma	guinea pig	Egeberg and Jensen [89]
Spontaneous leukemia, transplantable	lymphatic (L 1210)	mouse	Bekesi et al. [28]; Lefever et al. [208]; Killion [181]; Kollmorgen et al. [196]
Virus induced	mammary adenocarcinoma	mouse	Simmons and Rios [385, 388]
Virus-induced leukemia	AKR leukemia	mouse	Bekesi and Holland [30]
Spontaneous, non-transplantable partially metastasizing	mammary tumor (various histological types)	dog	Sedlacek et al. [352, 359]

Table XIII. (continuation)

Type of tumor	Histology	Type of animal	References
(b)			
Chemically induced non-metastasizing	fibrosarcoma (various types)	mouse	Spence et al. [397]; Wilson et al. [457]
	fibrosarcoma; hepatoma; mammary carcinoma	rat	Pimm et al. [271]
	glioma	mouse	Albright et al. [7]
	E 14 lymphoma	mouse	Ghose et al. [113]
Unknown cause transplantable, partially metastasizing	B 16 melanoma	mouse	Jamieson [166]; Froese et al. [107]^a
	Lewis lung adenocarcinoma	mouse	Sedlacek et al. [360]
	epithelioma, fibrosarcoma	rat	Pimm et al. [271]
Virus induced	polyoma, virus-induced adenocarcinoma	mouse	Porwit-Bobr et al. [274]
Spontaneous leukemia, transplantable	lymphatic (L 1210)	mouse	Killion [181]
Virus-induced leukemia	Rauscher leukemia	mouse	Barinskii and Kobrinskii [24]
	AKR leukemia	mouse	Mathé et al. [223]; Doré et al. [82]
Spontaneous, non-transplantable, partially metastasizing	mammary tumor (various histological types)	dog	Sedlacek et al. [352]

[a] Tumor growth acceleration after injection of VCN-treated cells

VCN-treated tumor cells without, however, itself influencing the growth of the solid tumor [386, 387]. On the other hand, additional suitable chemotherapy proved to be necessary in the immunotherapy of experimental leukemias and lymphomas [28–31, 69, 153, 196]. Whereas, for instance, in L1210 leukemia only approximately 10% of the animals could be cured by an optimal chemotherapy, additional immunotherapy with VCN-treated L1210 tumor cells on the 4th day led to survival of

20–55% of the animals and to a lengthening of the survival time [29]. Similar results were found in virus-induced leukemia in AKR mice. Here, even VCN-treated allogenous leukemia cells (E_2G leukemia cells) induced by the same virus were therapeutically equally effective as the syngenous tumor cells, which would point to cross-reacting antigens [29, 69, 154]. All in all, immunotherapy with VCN-treated inactivated tumor cells proved effective in a number of both chemically induced and spontaneously arising metastasizing or virus-induced transplantation tumors.

This effectiveness was dependent on the expression of membrane-bound glycoproteins, since treating the cells with galactosidase [181] or with protease [67, 68, 181] removed their immunogenicity and therapeutic effect completely.

There have, however, been reports on the ineffectiveness of immunotherapy with neuraminidase-treated tumor cells, especially in tumor models for which successes had been seen by the same or other investigators. In some investigations, even a more rapid tumor growth was seen as a response to the specific immunotherapy (see table XIII).

Wilson et al. [457] were the first to attempt to explain these different and sometimes contradictory findings: 24 h after surgical removal of the primary tumor transplant, various amounts ($1 \times 10^3 - 5 \times 10^5$) of inactivated tumor cells, partly after VCN treatment, were injected s. c. into metastasizing, methylcholanthrene-induced sarcoma, and the number of lung metastases in the mice recorded at certain intervals. Surprisingly, it was always only a certain number of VCN-treated tumor cells which proved to be therapeutically effective. Lower or higher amounts of cells were ineffective or even led to an increase in lung metastases.

In contrast, in our experiments with 3-LL-AC in the same strain of mice (C57B16/J) and in a similar animal model as used by Wilson et al. [457], the metastasis growth after primary tumor amputation could not be influenced either by low ($1 \times 10^4 - 10^5$) or by high (10^6; 2×10^7; 10^8) numbers of inactivated VCN-treated tumor cells [360, 362]. Thus, the contradictory results in the immunotherapy of various experimental rat and mouse tumors or its ineffectiveness cannot be simply explained by a cell-dose dependency. The specific sensitivity of a tumor for immunotherapy and, in addition, a possible dose-dependent adjuvant activity of cell membrane bound VCN have to be considered as further factors [371].

Most of the experiments on specific tumor immunotherapy with VCN-treated inactivated tumor cells were carried out in transplantation tumors. It is well known from the experience with cytostatics and non-antigen specific immunostimulators that the results in these models cannot be transferred to the clinical situation [363].

Considering this lack of clinical predictability, we also performed experiments with VCN-treated inactivated tumor cells in autochthonous spontaneous mammary tumors of dogs [352, 359]. This tumor is relatively frequent, clinically easily accessible, and, despite all the differences, it has certain similarities with human breast cancer. The most important similarities are a similar hormone and age dependency of development, certain prognostic factors (dependency on tumor size and infiltrative growth) [357], and the appearance of the so-called Thomsen-Friedenreich antigen [133, 362, 365]. Thus, therapeutic successes in dogs may provide much more information for the clinical situation in humans than results obtained from experiments with conventional mouse transplantation tumor systems.

For immunotherapy experiments, dogs were selected which had at least two mammary tumors. One tumor was surgically removed and the second tumor left in the animal for control purposes. On the day when the tumor was removed and on the following day, 1×10^6, 1×10^7 or 5×10^7 autochthonous (isolated from the removed tumor) VCN-treated inactivated tumor cells were injected subcutaneously into the neck of the dogs. For control purposes, a further group was simultaneously given 1×10^7 autochthonous inactivated tumor cells, not treated with VCN. Whole tumor cells and a total number of 2×10^6, 2×10^7 or 1×10^8 tumor cells per dog were chosen; firstly, since intact cells and this cell dosis range were able to provoke the best immune reaction in various tumor systems when administered intradermally or subcutaneously, either once or repeatedly, measured according to the protection rate against a subsequent injection of living tumor cells [16, 162, 209, 212, 239, 275–277]; and secondly, because this number of VCN-treated tumor cells revealed to be therapeutically effective at least in a part of the transplantation tumors in mice.

Similarly, as has been found in transplantation tumors [457] and also in the spontaneous mammary tumors of the dog, the therapeutic effectiveness of VCN-treated, autologous, inactivated tumor cells depended to a very large degree on the applied cell dose. Astonishingly, 2×10^7 VCN-treated tumor cells led to a clear regression of the mammary tumor remaining in the animal and to a prevention of metastasation or late progression in the majority of animals. This therapeutic effect was statistically significantly different from the control group (injection of 2×10^7 tumor cells, not treated with VCN) and could be found in three independent studies [352, 358, 359].

In contrast, injection of an amount of VCN-treated inactivated tumor cells 5 times larger than the therapeutically effective one caused accelerated progression of residual tumor mass in all cases, and all the dogs with malignant primary tumors died due to metastases within a year. On the other hand, after injection of 2×10^6 VCN-treated inactivated tumor cells, only a temporary regression with subsequent progression of the residual tumor could be seen, and there were no indications that repeating the injection twice would have prevented or influenced the subsequent progression.

Whether or not factors other than the number of tumor cells additionally influenced the results cannot be clearly said, due to the relatively small number of animals per group and the heterogeneity between the groups with respect to differences in ages and control tumor volumes as well as histological types of primary tumors.

There was no clear correlation between the therapeutic success and the volume of the residual tumor mass. Thus, tumors with an initial volume of under 1 cm^3 had the same rates of progression and regression as those with a volume of 2 or 3 cm^3. Furthermore, no relationship could be seen between the amount of NeuAc separated from the cell membrane and the age of the dog, the age of the tumor, or primary tumor histology on the one hand and the therapeutic or otherwise effect of immunotherapy on the other.

Due to the type of this tumor model, it is difficult to draw clear conclusions. Thus far, it is unclear whether multiple dog mammary tumors develop independently of one another or whether they influence one another or have common causes. As yet, there have been no indications of virus genesis [93], but clear ones for a common dependency on endocrinological factors [14, 71, 83, 218, 242, 328, 341, 430]. Nevertheless, the relationship of the primary tumor to the control tumor is unclear. Histological investigations show either a complete homogenicity in malignancy of multiple mammary tumors [198] or a 23% heterogenicity in multiple tumors [96]. In our own studies [352, 359], the histological diagnosis of the control tumor corresponded to that of the primary tumor in about 80% of dogs.

In view of the lack of clarity regarding the degree and type of common factors in dog multiple mammary tumors, it is surprising that an injection of VCN-treated tumor cells from the primary tumor is able to influence the growth of a secondary mammary tumor, the control tumor. If a tumor-specific influence is assumed, both tumors have to have common antigens against which an enhanced specific immune reaction is provoked during therapy. For logistic reasons (lack of autologous tumor material and too small a cell harvest during production of individual tumor cell suspensions), specific immune reactions (antibodies and cellular immune reactions against autologous tumor material) subsequent to application of VCN-treated tumor cells could not be recorded by us.

Histological investigations showed that tumor tissue was demonstrable in the control tumor despite a long lasting and stable regression. Given a tumor-specific immune reaction, the cause of this stable regression would have to be assumed to be a stable balance, the details of which are unknown, between tumor cell proliferation and immunological resistance reaction.

All in all, injection of inactivated VCN-treated autologous tumor cells proved to be therapeutically effective in spontaneous mammary tumors in dogs. The therapy with VCN-treated inactivated tumor cells (no death due to metastasis within about 3 years) was superior not only to that in the control group, i. e., treatment with inactivated tumor cells (resulting in a death rate of about 25% caused by the tumor), but it also appears to be better than radical operation [430] (death rate of 24% within about 3 years), despite the fact that the control tumors remained in the dogs.

Despite this clear therapeutic effect, tumor therapy with VCN treated tumor cells must be considered in a critical light. The effective cell dose is too close to the only temporarily-effective or even tumor-growth-accelerating number of cells. The immunopharmacological mechanism causing these differing results is unclear; it might be due to T-suppressor cells which are stimulated to different degrees depending on the tumor cell dose [54, 192], or to antibodies which immunosuppress via an immune complex reaction [348, 349], or to non-antigen specific mechanisms such as activation of macrophages with suppressive function [368]. The role of those mechanisms, however, in immune surveillance of the tumor growth, is still not clearly understood. Thus, it is impossible to establish an immunotherapeutically effective cell dose via registration and quantification of these mechanisms. Since the possibly decisive imunological mechanisms might be different from tumor to tumor and, moreover, appear to depend on the constitution and condition of the tumor bearer, the injection of a certain amount of VCN-treated tumor cells can be accompanied by the risk of thereby accelerating tumor growth. Thus, an uncritical use of VCN-treated tumor cells in humans cannot be advised.

The bell-shaped curve dose-dependent antitumoral effectiveness of VCN-treated tumor cells has parallels in non-specific immunomodulators such as BCG, vitamin E, or zymosan [42], or muramyl dipeptides [468]. Here too, a V- or W-shaped dose-effect curve can be seen, and the question arises as to which method can provide the optimally effective dose with respect to individual variability of immune reactivity. At present, there is no ready answer to this question.

One step towards a solution of the cell-dose problem in tumor therapy with VCN-treated tumor cells has been ventured by Bekesi and Holland [29, 32, 33, 154, 155]. In their randomized prospective study, they selected the dose for immunotherapy of acute myeloid leukemia (AML) with the aid of the DTH reaction. In the course of an orientating study, they injected patients with various amounts of allogenous VCN-treated AML cells after successful induction chemotherapy. Of the cell dose which caused the maximal DTH reaction it was assumed that it was also the most effective dose therapeutically, and this dose was used for immunotherapy in their successful studies on specific immunotherapy of AML with VCN-treated AML cells.

Combination of VCN-Treated Tumor Cells and Non-Specific Immunostimulants

In order to increase the therapeutic use of specific tumor therapy with VCN-treated cells, animals were additionally treated with BCG as a non-specific immunostimulator [197, 380, 388]. In a series of experiments in mice (methylcholanthrene-induced fibrosarcoma, L1210 leukemia), this combination was found to be superior to the respective individual treatment, and this fact led to combination experiments (VCN-treated cells and BCG or methanol-extracted residue of BCG) in patients with melanomas [389] or with AML [31–34]. The results, however, did not differ from those of the control group without any sort of immunotherapy and were surprisingly worse than after single specific immunotherapy with VCN-treated cells (AML). One explanation for this could be provided by our adjuvant experiments [356]: treatment with living BCG germs as a non-specific immunostimulant increased the immunological reaction against a subsequent injection of SRBC as a non-related antigen. This BCG-induced, increased non-specific immune reaction against the antigen, however, could not further be specifically enhanced by addition of the adjuvant VCN to the SRBC. In case the antigen represents a tumor-associated antigen, it might consequently be concluded that the combination of specific (VCN-treated tumor cells) and non-specific (BCG) tumor immunotherapy does not seem likely to offer any advantages against single immunostimulation. Indeed, the already mentioned negative clinical results and the equally negative results of the subsequent investigations in mice on the effect of BCG on tumor immunotherapy with VCN-treated tumor cells [397] and, likewise, the data on the substractive effect of BCG in combination with tumor cells [182] seem to confirm our assumption.

Nevertheless, it remains to be clarified whether or not with a more detailed knowledge of the immunopharmacology of non-specific immunostimulants, for instance, those of the recent generation, a combination of these substances with specific immunotherapy would have additive or even synergistic therapeutic effects on tumor diseases.

Administration of Mixtures of Tumor Cells and VCN

In experiments to increase the immunogenicity of tumor cells, as well as in immunotherapy of experimental tumors using VCN-treated tumor cells, the observation was made [30, 373, 380, 381] that there was a relationship between therapeutic success and the amount of VCN used for cell treatment. An increase in enzyme concentration, 2 or 3 times higher than the optimal concentration, lessened the therapeutic effect. The conception of NeuAc separation from the tumor cell membrane by VCN cannot be the explanation of this phenomenon, neither can a cytotoxic effect of VCN [159, 299, 313, 316, 317]. However, our studies on cell membrane binding [213, 350, 371] and the results we obtained on the adjuvant activity of VCN [170, 191, 356] point to a possible explanation. The adjuvant effect of VCN showed a single peak dose-effect curve similar to the dose-effect curve for VCN-treated tumor cells in tumor immunotherapy in dog mammary carcinoma. Since our investigations showed that cell membrane bound VCN continues to be enzymatically active, it seemed possible that the increased cell immunogenicity after VCN treatment and the therapeutic effect of VCN-treated tumor cells were caused by the adjuvant activity of cell membrane bound VCN. Consequently, the amounts of VCN which were injected with a therapeutically effective number (1×10^7) of VCN-treated tumor cells per injection site in our therapy studies in dogs suffering from mammary carcinoma and humans with AML (2×10^8) [410] were assessed using radioactively labeled VCN [213]. Surprisingly, approximately similarly high VCN values were found (Lüben, personal communication in [371]), although in the AML study about 20 times more cells were injected per injection site than in the study on dog mammary tumors. Thus, the effectiveness of tumor immunotherapy with VCN-treated tumor cells not only seems to depend on the number of tumor cells but also on the amount of VCN which, adhering to the surface of the tumor cells, is injected with the tumor cells.

Thus, it was logical to try to optimize tumor therapy by admixtures of VCN to inactivated tumor cells and by administering these mixtures instead of using VCN-treated tumor cells.

Experimental investigations on intratumoral injection of VCN have already been presented by a number of authors (table XIV). Both regression and ineffectiveness or accelerated growth of the injected tumor [38] or of the second tumor [384, 396] had been found. Tumor regressions were found to be dose dependent since only intratumoral injections of larger amounts of VCN were therapeutically effective. As a mechanism, an in-vivo development of VCN-treated tumor cells after intratumoral injection of VCN was assumed. Since their number was unknown and can be variable, this could explain (also in view of the small therapeutic spectrum of VCN-treated tumor cells) the contradictory therapeutic results after intratumoral application of VCN.

Alternatively, we attempted [360] to influence the growth of metastases after removal of the primary tumor transplant of the 3-LL-AC by subcutaneous or intradermal injection of tumor cells mixed with various amounts of VCN and aministered 1 or 6 times. However, the death rate or survival time of the mice could not be influenced either by relatively low (down to 1×10^6) or by high (up to 2×10^8) numbers of inactivated tumor cells mixed with small (10 mU) or larger (100 mU) amounts of VCN.

Thus, in the therapy of this tumor, individual mixtures of tumor cells and VCN were equally ineffective as VCN-treated tumor cells or tumor cells alone, although in the protection experiments already dis-

Table XIV. The therapeutic effectivity of intratumorally injected VCN

	Type of tumor	Histology	Species	References
Regression	chemically induced	fibrosarcoma	mouse	Gautam and Aikat [112]; Sparks and Breeding [396]
		Ehrlich ascites	mouse	Mobley et al. [236]
	spontaneous, metastasizing, transplantable	Lewis lung adenocarcinoma	mouse	Alley and Snodgrass [9, 10]
	virus-induced	mammary carcinoma		Simmons et al. [384]
No effect	spontaneous, metastasizing, transplanting	skin carcinoma	rat	Binder et al. [38]
Acceleration	spontaneous, metastasizing, transplantable	skin carcinoma	rat	Binder et al. [415]
	virus-induced	mammary carcinoma	mouse	Simmons et al. [416][a]
	chemically induced	fibrosarcoma	mouse	Sparks[a] and Breeding [417]

[a] Acceleration of secondary tumor growth

ussed, mixtures of 10^6 or 5×10^6 tumor cells and 100 mU VCN were effective at least once and thus superior to all other single combinations or tumor cell preparations (table XV). The reason for this therapy failure may lie in the falsely chosen cell and VCN mixtures not being able to successfully interfere with the immune system to achieve a therapeutic effect.

Chessboard Vaccination

The development of the chessboard vaccination has its origin in the clinical experience gathered by Bekesi and Holland [32, 33], who injected patients intradermally with different amounts of VCN-treated tumor cells and were able to show that the amount of VCN-treated tumor cells which caused a maximal DTH reaction in AML patients was also the one which was effective in maintenance therapy of AML.

Table XV. Specific immunotherapy of experimental transplantation tumors

Immunotherapy application of	Appl. site	Success rate of immunotherapy[a]			
		lung adeno-carcinoma (3-LL-AC)	melanoma (B-16-M)	mammary adeno-carcinoma (R 3230-AC)	
Cells					
(1) 10^6; 2×10^7 or 10^8 M-TC	s.c.	∅	n.d.	n.d.	
	i.d.	∅	n.d.	n.d.	
(2) 10^6; 2×10^7 or 2×10^8 M-TC-VCN	s.c.	∅	n.d.	n.d.	
	i.d.	∅	n.d.	n.d.	
Single mixtures					
(3) 10^6; 2×10^6; 10^7; 2×10^7; 10^8 or 2×10^8 M-TC + 10; 25; 50 or 100 mU VCN	s.c.	∅	n.d.	n.d.	
	i.d.	∅	n.d.	n.d.	
Combinations of mixtures					
(4) 10^5; 10^6 and 10^7 M-TC	i.d.	∅	∅	∅	
(5) 10^5; 10^6 and 10^7 M-TC + 100 mU VCN each	i.d.	∅	∅	∅	
(6) chessboard: 10^5; 10^6; 10^7 M-TC; each mixed with 10; 50; 100 mU VCN	i.d.	20	30	30	
(7) chessboard with heat-inactivated VCN	i.d.	∅	∅	20	
Immunotherapy and chemotherapy					
(8) CPPD	i.v.	40	20	10	
(9) CPPD + IT [5]	i.v./i.d.	70	10	n.d.	
(10) CPPD + (2×IT [5])	i.v./i.d.	100	10	20	
(11) CPPD + chessboard [6]	i.v./i.d.	n.d.	10	0	

[a] % reduction in mortality due to metastazation (10 animals per group). Immunotherapy was performed 1 day after resection of a primary tumor graft. Death in case of metastasis was recorded.

M-TC: Mitomycin-treated cells (100 μg/10^7 cells; 60 min; 37°C) of the respective tumor
M-TC-VCN: VCN-treated M-TC (100 mU VCN/5×10^7 cells/ml PBS)
CPPD: 100 mg/kg (mice) resp. 20 mg/kg (rat) on day of tumor resection
n.d.: not done
According to Sedlacek et al. [365]; Sedlacek et al. [360]

On the basis of these results, we tried to find out which mixture caused the strongest DTH reaction by intradermally injecting various mixtures with increasing numbers (10^5, 10^6, 10^7, 10^8) of inactivated tumor cells and with increasing amounts of VCN (10, 50, 100 mU). This injection procedure was called chessboard vaccination [371]. Dogs with mammary tumors (multiple tumors) were used as a model since DTH reaction can be more easily recorded than in the mouse or rat, and we already had experience of successful immunotherapy with VCN-treated tumor cells [352, 359].

In the course of a pilot study [359] on the day of resection of one of the multiple mammary tumors, all the various mixtures of autologous tumor cells and VCN were simultaneously, and separated from each other, injected intradermally into the dog. Immediately afterwards, and 24 and 48 h later, DTH was determined by measuring skin fold thickness. Since, in accordance with the results obtained in our preceding study with VCN-treated inactivated tumor cells, it was to be feared that the high number of cells administered might cause an acceleration in tumor growth, the animals were subjected to continuous clinical examination and, in particular, the volume of the tumor remaining in the animal was determined. Moreover, for comparison, dogs in a second control group were treated with subcutaneous injection of a total of 2×10^7 inactivated VCN-treated autologous tumor cells.

Measurement of skin reaction did show in all dogs a DTH reaction against autologous tumor cells which had its maximum in the case of mixtures consisting of a large number of tumor cells and a small amount of VCN or a small number of tumor cells and a larger amount of VCN. On the other hand, a large number of dogs showed tumor regression after intradermal injection of such a chessboard-like combination of mixtures of tumor cells and VCN. Judging by the results obtained in the positive control group, the therapeutic effectiveness of chessboard vaccination was not worse than that of subcutaneous injection of 2×10^7 inactivated VCN-treated tumor cells [359].

In a subsequent prospective randomized and controlled study, the specificity and the role of enzymatically active VCN in chessboard vaccination were investigated [133, 362]. Dogs with multiple mammary tumors, randomized in 3 different groups, were treated either with autologous tumor cells mixed with active VCN, with autologous tumor cells mixed with heat-inactivated VCN, or with autologous erythrocytes mixed with active VCN.

After an observation period of approximately 1 year, constant or, in part, temporary regressions could only be seen in groups which had received chessboard vaccination with autologous tumor cells and active VCN. Moreover, no metastases appeared in this group. In contrast, in both control groups a number of dogs died of metastases and in no dog could a regression of the tumor remaining in the body be observed (table XVI).

In view of these significant differences, it would seem logical to conclude that chessboard vaccination is a method of specific tumor therapy which depends for its therapeutic effectiveness on the use of tumor cells and enzymatically active VCN. It is questionable whether other factors influenced the results of this study similarly as in the studies with VCN-treated tumor cells [352, 359]. There were no obvious dependencies on the age of the dogs, the age of the tumor, volume of the control tumor, histological diagnosis of the primary tumor or the control tumor, and/or the amount of the Thomsen-Friedenreich antigen, evaluated according to peanut lectin binding to sections of the primary tumor. As in the other studies on dog mammary tumors, here, too, the question remains unanswered as to the relationship between the primary and control tumor.

The question automatically arose whether chessboard vaccination is only effective in autochthonous dog mammary tumors, or whether it can also be shown to be effective in experimental transplantation tumors. To this aim, we carried out investigations in 3 different metastasizing tumor models: mouse 3-LL-AC, mouse B16 melanoma, and rat R 3230-AC, whereby chessboard vaccination took place on the day of operation after removal of the primary tumor transplant. The death rate due to metastases and the survival times of the animals dying due to metastases were recorded [360].

Subcutaneous or intradermal injection of different numbers of VCN-treated inactivated tumor cells or of single mixtures consisting of inactivated tumor cells (10^5, 10^6, 10^7, or 10^8 cells) and 10, 25, 50, or 100 mU VCN had no therapeutical influence on growth of the tumor in 3-LL-AC. Similarly ineffective, both in 3-LL-AC and R-3230-AC, was a simultaneous injection of various numbers of inactivated tumor cells alone (10^5, 10^6 and 10^7 cells) or each mixed with a constant amount of 100 mU VCN. In contrast, chessboard vaccination (with increasing numbers of tumor cells each mixed with increasing amounts of VCN, i.e., in the form developed and carried out with dogs) clearly reduced the number

Table XVI. Specific immunotherapy of spontaneous mammary tumors in dogs

Study observation period	Immunotherapy	No. of animals	Appl. site	Results (growth of the residual tumor mass examined on days 0; 28; 56; 112; 224; 224)				
				regression	transient regression	no change	pro-gression	death due to tumor
I (3 years)	2×10^7 M-TC	12	s.c.	2	1	2	7	5
	2×10^7 M-TC-VCN	15	s.c.	13	0	1	1	2
	1×10^8 M-TC-VCN	8	s.c.	0	0	0	8	6
II/III (1 year)	2×10^7 M-TC-VCN	14	s.c.	7	3	1	3	0
	2×10^6 M-TC-VCN	15	s.c.	0	11	1	3	2
	chessboard (M-TC + VCN)	10	i.d.	7	0	1	1	0
IV (1 year)	chessboards M-TC + VCN	23	i.d.	2	4	0	17	0
	M-TC + VCN 100°C/10 min	21	i.d.	0	0	0	21	3
	Ery + VCN	27	i.d.	0	0	0	27	2

M-TC: Mitomycin-treated (25 μg/10^6 cells; 60 min 37°C) autologous tumor cells
M-TC-VCN: VCN-treated M-TC (100 mU VCN/5×10^7 cells/ml PBS containing 0.9 mmol $CaCl^2$/l)
Chessboard vaccination: Intradermal application of a chessboard-like series of mixtures of 10^5; 10^6; 10^7; 10^8 M-TC each mixed with 0; 10; 50; 100 mU of VCN
Immunotherapy: performed on the day of and on the day after excision of a part of the tumor mass
Sedlacek et al. [133, 352, 360, 362, 365]

of mortalities due to metastases from 100% to between 50% and 80% in 3 independent experiments with 3-LL-AC. Chessboard vaccination carried out as a control with heat-inactivated VCN or PBS was ineffective (table XV).

In contrast, chessboard vaccination had no significant or reproducible (R-3230-AC) therapeutic effect on B16 melanoma and R-3230-AC, in spite of the fact that those tumors had to be regarded at least as weakly immunogenic as the 3-LL-AC [31, 51, 53, 144, 145, 200, 222, 274].

Thus, in conclusion, chessboard vaccination was shown to be at least as therapeutically effective on tumor diseases as, or superior to, an injection of VCN-treated tumor cells or single mixtures of tumor cells and VCN. This therapeutic effect, however, is restricted to certain tumor diseases and to certain models. It is completely unknown why other tumor diseases are resistant to this form of therapy.

A simple explanation might be that tumor cell proliferation is too strong to be overcome by immunotherapy. Combinations of substances which inhibit tumor proliferation, administered together with specific immunotherapy, would thus have the chance of more therapeutic success provided the cytostatic therapy used is not immunosuppressive and allows specific immunostimulation. The effectiveness of this combined chemo- and immunotherapy could already be clearly demonstrated in rapidly proliferating L1210 leukemia [29]. A cytostatic drug (MCCNU) was chosen for cytoreduction at a dose of which the effect on the L1210 cells was suboptimal; lymphocyte proliferation (stimulated by PHA) was, however, not essentially impaired. Immunotherapy with VCN-treated tumor cells was carried out 4 days after chemotherapy. Whereas approximately 10% of the animals were cured by chemotherapy alone, 20 to 55% survived after chemoimmunotherapy. Pincus et al. [272] achieved similar results in the same tumor model with MCCNU combined with VCN-treated cell membrane preparations from L1210 leukemia cells.

Thus, the question was whether the therapeutic results of chessboard vaccination could be accordingly improved by additional chemotherapy. For such experiments, cyclophosphamide was selected, because it is cytostatic for 3-LL-AC. Moreover, in a dose range of 5 mg/kg – 200 mg/kg, with an optimal dose of 100 mg/kg [306], cyclophosphamide affects mainly tachytrophic cells cytostatically [283, 345, 429] as, for instance, T-suppressor cells [80]. Pretreatment with cyclophosphamide or simultaneous administration of this cytostatic drug and an immunogen thus leads to an increase in the (T-suppressor lymphocyte regulated) immune reaction and that of DTH against several immunogens [18, 221, 427, 428], including SRBC [116, 180, 203], moreover, to an increase in cytotoxic T-lymphocytes [118, 306] and/or in the antibody response as well. Some other cytostatic drugs work in a way similar to cyclophosphamide, either stimulating the immune response or suppressing it according to the dose and the relationship in time to the administration of the immunogen [47, 398, 435].

In our experiments mice were treated with 100 mg/kg cyclophosphamide on the day of and the day after amputation of the primary tumor (3-LL-AC) transplant and subsequently intradermally injected with a series of different numbers (10^5, 10^6, 10^7) of inactivated tumor cells, each mixed with a constant amount (100 mU) of VCN. Whereas this type of immunization (unlike the chessboard vaccination which includes variable amounts of CNV) had no effect in itself on the subsequent growth of metastases of 3-LL-AC, cyclophosphamide, when administered alone, reduced metastase-induced mortality from 100% to 60%. Surprisingly, the combination of chemotherapy and specific immunotherapy led to a further reduction of the mortality rate to 30%, and a repetition of specific immunotherapy 9 days after administration of the combination led to recovery of all animals [362].

Thus, our experiment clearly shows that the combination of suitable chemotherapy and a specific immunotherapy (ineffective on its own) can lead to an improvement of the therapeutic results. It is unclear whether this increased effect is due to a chemotherapeutically induced inhibition of tumor cell proliferation alone or is also due to cyclophosphamide-mediated inhibition of T-suppressor cells which are stimulated by 3-LL-AC [222, 391, 422]. This inhibition of T-suppressor cells by cyclophosphamide could make an ineffective type of specific immunostimulation now therapeutically effective. As the chessboard vaccination alone is therapeutically effective in 3-LL-AC, too, the conclusion might be drawn that chessboard vaccination by itself abrogates stimulation of T-suppressor cells. Neither growth of B16 melanoma in mice nor of R 3230-AC in rats could significantly or reproducibly be impaired by the selected dose of cyclophosphamide or by chessboard vaccination or by a combination of both treatments [362]. Thus, suppression of suppressive T-cells by cyclophosphamide does not seem to be the only reason for the success of combination treatment. Impairment of tumor growth by the selected cytostatic drug and a special susceptibility of the respective tumor type for immunotherapy also has to be taken into consideration.

It is unclear why chessboard vaccination is therapeutically more effective than a single injection of a certain selected number of cells treated or mixed with VCN. It is still an unproven assumption that an antigen-specific stimulation of the immune reaction due to application of tumor cells and VCN is the reason for the therapeutic success. Based on this assumption, however, it might be speculated that chessboard vaccination is better suited to the variable and individual reactive situation of the immune system of the receiver. In view of this variability, the probability of immune stimulation by a certain combination of tumor cells and VCN might be greater with the chessboard vaccination than with an injection of a single fixed number of tumor cells, whether VCN-treated or mixed with VCN.

In the course of our investigations with dogs it was striking that chessboard vaccination did not provoke an accelerated tumor growth in any dog [359, 360, 362], even though cell amounts were injected which, when administered alone and subcutaneously, had caused premature death in dogs due to metastases [352, 359]. A comparison of the two therapy experiments is only possible to a limited extent due to the different sites of application. However, in order to explain the phenomenon, we assumed [371] that in combinations of stimulating and suppressing cell doses, as in chessboard vaccination, the stimulating tumor-cell numbers may have a stronger effect than all the other cell doses. This contradicts previous experience, which has shown that when applied simultaneously suppressogenic determinants dominate over immunogenic determinants [241].

The use of autologous tumor material for chessboard vaccination limits its use in tumor therapy to cases in which enough tumor cells can be isolated from the surgically removed tumor tissue. Thus, technical facilities for tumor cell isolation are a prerequisite. A simplification of the therapy and, thus, accessibility to clinical practice would only be possible if allogenous culture cells could be used instead of freshly harvested autologous tumor cells. Bekesi et al. [29, 30, 69] have already shown in the case of AKR-leukemia that allogenous tumor cells can also be effective therapeutically if there are common factors in the tumor-associated structures. First experiments in dogs showed that the application of allogeneic cultured mammary tumor cells either treated with VCN or mixed with VCN and injected as chessboard vaccination might also have a therapeutic effect [359, 371].

Since tumor cells have organ-specific, individual-specific, and species-specific antigens besides the possibly tumor-associated antigens, the question arose as to whether or not an enhancement of these antigens with VCN in the course of tumor therapy with VCN-treated tumor cells or, possibly to a greater extent, in the course of chessboard vaccination could lead to autoimmune reactions. Clinical and serological investigations, including autoantibodies against cell nuclei, mitochondria, and the smooth muscular system as well as blood picture provided no indication of an enhancement or induction of an autoimmune reaction in dogs either after administration of autologous tumor cells or autologous erythrocytes [133, 362].

VCN itself was also found to be locally tolerable and non-immunogenic within chessboard vaccination. Antibodies against VCN (probably cross-reactive, as we have already been able to observe in humans) [171]) did not increase after chessboard vaccination nor did anti-VCN antibodies develop subsequent to chessboard vaccination [133, 365]. These observations are in agreement with the data concerning VCN toxicity and immunogenicity [310].

All in all it can be concluded that chessboard vaccination appears to be a safe and effective procedure for specific therapy of selected tumors.

The Role of Thomsen-Friedenreich Antigen

Of course, the question arises as to which antigen structures on the surface of tumor cells are responsible for the therapeutic effect and whether lack of those antigens might explain ineffectiveness in certain tumor models of immunotherapy with tumor cells and VCN. Thomsen-Friedenreich (T-) antigen could be shown by us on the mammary carcinoma tissue and in mammary tumor cells in dogs [133, 362, 365], in the B16 melanoma, the R 3230-AC, and in LL-AC. After VCN treatment, Thomsen-Friedenreich antigens exposed by those tumor cells were enhanced, but not clearly in LL-AC. After VCN treatment, dog erythrocytes and sheep erythrocytes (SRBC) also expose Thomsen-Friedenreich antigen [133, 365, 439]. Those investigations have been performed with the aid of peanut lectin binding which, as already discussed, reacts with the immune dominant carbohydrate fragment as immune dominant T-antigen structure [174, 278, 431–433], and with the aid of human sera containing antibodies against the T-antigen, by immunofluorescent histological methods, and by means of agglutination investigations.

Despite the fact that dog erythrocytes and dog mammary tumor cells expressed the Thomsen-Friedenreich antigen in similar ways after VCN treatment, autologous dog erythrocytes, in contrast to autologous tumor cells, proved to be therapeutically ineffective in chessboard vaccination in dog mammary carcinoma. An alteration or the appearance of an antibody titer against T-antigen was not induced by chessboard vaccination. Additionally, the number of primary or control tumors which were able to bind peanut lectin, with or without pretreatment of the tissue sections with VCN, was the same in all treatment groups [133, 365].

In the case of transplantation tumors, a therapeutic success could only be seen with chessboard vaccination in LL-AC which, unlike B16 melanoma and R 3230-AC, expresses only a small amount of Thomsen-Friedenreich antigen, the degree of which could not be enhanced by VCN.

A further question consisted of demonstrating an increase in antigen-specific immune reaction due to a therapy comprising tumor cells and VCN. For logistic reasons, only the DTH reaction against autologous tumor material could be investigated by us in dogs [133]. Comparable investigations were carried out in mouse LL-AC [365]. The DTH reaction in mice showed no clear differences between chessboard vaccination with active or inactive VCN. In the case of dog mammary carcinoma, a difference could not be found in the absolute or relative values of the maximal increases in skin thickness between the control groups and the test groups [64, 133, 365]. Thus, against autologous erythrocytes mixed with VCN, a DTH reaction could be observed which was of the same magnitude as the DTH reaction against autologous tumor cells mixed with active VCN or inactivated VCN. In view of the fact that dog erythrocytes express, similarly to canine-treated and untreated mammary tumor cells, the Thomsen-Friedenreich antigen after VCN treatment, it cannot be excluded that the recorded DTH reaction may have been directed against Thomsen-Friedenreich antigen. As chessboard vaccination with autologous tumor cells and VCN was only therapeutically effective, we may additionally conclude that the DTH reaction against a cell exposing Thomsen-Friedenreich antigen is not necessarily an indication of a therapeutic effect the intradermal application of this cell might have on tumor diseases. Whether, as supposed for humans [405–408], the DTH reaction against the Thomsen-Friedenreich antigen is suitable for the diagnosis of selected tumors has not yet been elaborated in experimental tumor models.

All in all, on the basis of the lack of correlation between the therapeutic effect of chessboard vaccination on the one hand and the demonstration of the Thomsen-Friedenreich antigen on the other, we may draw the conclusion that the Thomsen-Friedenreich antigen has no decisive influence during tumor therapy with VCN and tumor cells either in dog mammary tumor or in the transplantation tumors examined [365].

Clinical Studies

VCN-Treated Tumor Cells

The first clinical experiments to achieve regression of solid tumors by means of subcutaneous injection of VCN-treated inactivated autologous tumor cells were carried out by Seigler et al. [366] and by Rosato et al. [311, 312]. Seigler et al. treated 7 patients with metastasizing or relapsing melanoma. In 6 of these 7 patients they observed a complete regression of the tumor mass. However, since either BCG alone or in combination with autologous lymphocytes sensitized in vitro against autologous melanoma cells were administered at the same time as the VCN-treated tumor cells, the clinical findings could not clearly be attributed to one of the methods of treatment. Rosato et al. [311] injected intramuscularly into 25 patients with various tumors (mammary carcinoma, gastrointestinal carcinoma, carcinoma of the uterus and kidney) 5×10^6 VCN-treated autologous inactivated tumor cells each month for 6 months. In all the patients, the illness was at an advanced stage and resistant to other forms of treatment. Since the investigators found no promoting of the tumor disease by 'immunotherapy', no side-effects, and an increased lymphocyte-mediated cytotoxicity against various target cells in vitro, Rosato et al. [312] widened the scope of their study to 45 patients with various tumor illnesses.

Once more, they found both in vitro and clinical indications for an effect and at least no negative influences on the tumor illness and could confirm the good tolerance. Similar findings were made by Takita et al. [416] in patients with bronchial carcinoma. All in all, these first clinical investigations were not able to provide proof of the effectiveness of tumor immunotherapy with VCN; however, they did serve as an orientation point and as a basis for the planning of randomized, prospective, and adequately controlled studies.

Holland and Bekesi [154–156] carried out one of the first prospective, controlled, randomized studies with AML. A total of 175 patients in 2 studies were treated with maintenance chemotherapy according to the ALBG-protocol after successful induction chemotherapy with Ara-C and daunorubicin. One group of patients was additionally given VCN-treated allogenous AML cells, a third group was treated with VCN-treated AML cells plus methanol-extracted residue (MER) from BCG. Immunotherapy was carried out every 4 weeks and with a period of 2 weeks between it and the maintenance chemotherapy. Within this space of time, and due to preliminary experiments, a recovery from the chemotherapy-induced suppression of lymphocyte reactivity took place. The intradermal cell dose per injection site of 2×10^8 VCN-treated AML cells was selected on the basis of the DTH reaction. All in all, per immunization, 10^{10} blasts were injected over 48 injection sites. Maintenance chemotherapy was carried out over a period of 4 years and immunotherapy over 5 years, each at monthly intervals. An evaluation of 159 patients [34] after 5 years showed that in the control group with chemotherapy alone 9 out of 65 patients remained in remission; in the group with additional specific immunotherapy 24 out of 73 patients; and, surprisingly, in the group with the combination of specific and non-specific 'immunotherapy' in addition to chemotherapy, not one of the 21 patients. For patients with chemotherapy and specific 'immunotherapy' (especially for those over 45 years) there was a statistically significant lengthening of the survival time and the remission duration from 3–8.5 years. Parallel control investigations of lymphocyte proliferation (stimulated by mitogens (PHA or PWM) or by autologous AML cells) indicated a restoration of immunocompetence, quite unlike those patients with chemotherapy alone or chemotherapy combined with both specific and non-specific 'immunotherapy'.

Especially in the combination of methanol-extracted residue (MER) of BCG and VCN-treated AML cells, even less-than-average lymphocyte proliferation values were found so that a MER-induced activation of T-suppressor cells was assumed [34]. This could be an explanation for the lack of therapeutic effect in the combination of specific and non-specific 'immunotherapy' in contrast to specific 'immunotherapy' alone. A further explanation is offered by our adjuvant tests [356] which show that an additional specific immunostimulation is unsuccessful in an organism non-specifically immunostimulated with BCG.

In a third study, Bekesi et al. [34] tried to confirm the positive results gained in the 2 preceding studies. However, they improved the maintenance chemotherapy by including corticosteroids. A total of 92

Table XVII. Clinical studies on specific tumor therapy with VCN-treated tumor cells

Tumor (stage)	No. of pat.	Study protocol	Chemotherapy	Immunotherapy	Interim report	Results Final report	Authors
AML (No. 1)	14	random. 2 groups	Ara-C and DNR; Ara-C and TG/CP/DNR	I) — II) 10^{10} VCN-blasts (allog.) 48 injection sites; monthly		significant increase in survival time and remission time	Bekesi and Holland (Buffalo) [33, 34]
AML (No. 2)	53	random. 3 groups	Ara-C and DNR; Ara-C and TG /CP/DNR/CCNU	I) — II) VCN blasts (allog.) III) VCN blasts and MER (1 mg)		significant increase (II) compared to (I) in survival time and remission time MER (III) removes the therapeutical effect of (II)	Bekesi and Holland (New York) [33, 34]
AML (No. 3)	84	random. 2 groups	Ara-C and DNR; Ara-C/TG/VCR /DM/DNR	I) — II) VCN blasts (allog.)		no significant increase in survival time and remission time (II)	Bekesi and Holland (New York) [34]
AML	61	random. 2 groups	Ara-C and DNR (ADM) Ara-C and TG	I) — II) splenectomy III) 10^{10} VCN blasts (allog.) 48 injection sites; monthly	tendency towards increase in survival time and remission time (III)		Wiernik et al. (Baltimore) [364, 449]
AML		random. 2 groups	Ara-C and DNR and TG; Ara-C and DNR /TG/CP	I) — II) 10^{10} VCN blasts (allog.) 48 injection sites; monthly	significant increase in survival time and remission time after 2 courses of therapy (II)		Pielken et al. [270]; Urbanitz et al. (Münster) [438]

Clinical Studies

Diagnosis	N	Design	Chemotherapy	Treatment groups	Results	Reference
AML	26	random. 2 groups	Ara-C and DNR; COA/TRA/POM	I) — II) 5×10^8 VCN blasts (allog.) every 4 weeks	no difference from controls	Rühl et al. (Berlin) [322]
Melanoma	7	pilot study	varied	VCN cells (autolog.) and BCG	tumor regression in 6 patients	Seigler et al. [366]
Various solid tumors, treated	45	pilot study	varied	1×10^8 VCN cells (autolog.) monthly for 6 months	delay in tumor growth, increase in lymphocyte cytotoxicity	Rosato et al. (Philadelphia) [311, 312]
Melanoma stage II	22	random. 2 groups	none	I) — II) 2×10^8 VCN cells (autolog.) and BCG (3×10^8)	no difference	Simmons et al. (Minneapolis) [389]
Melanoma stage III	36	random. 2 groups	DTIC	I) — II) 2×10^8 VCN cells (autolog.) and BCG (3×10^8)	no difference	Simmons et al. (Minneapolis) [389]
Melanoma stages I and II	174	random. 4 groups	Me.-CCNU (group IV)	I) — II) BCG III) BCG and VCN cells (allog.) cultivated cells IV) Me.-CCNU	no difference	Terry et al. (Bethesda) [419]
Colon carcinoma	240	random. 3 groups		I) — II) VCN-treated autolog. cells and BCG III) BCG	started in 1980, no difference to controls	Gray et al. (Melbourne) [364, 449]

patients had been randomly distributed to either maintenance chemotherapy alone or combination of chemotherapy and immunotherapy. Up to now (i. e., 6 years after onset of this study), they could not find any statistically significant difference in remission duration and survival time between both groups. The speculation was that the inclusion of corticosteroids into the chemotherapy protocol might have prevented specific immunostimulation by VCN-treated AML cells.

Specific 'immunotherapy' of AML with VCN-treated AML cells, in accordance with the original program worked out by Bekesi et al. [34], is at present being tested in randomized controlled prospective studies by Wiernik et al. (cited in [449]) and by the group of Urbanitz [270, 437, 438]. However, maintenance chemotherapy [449] and induction chemotherapy [270] have been changed in comparison to the original protocol employed by Bekesi et al. [34]. Nevertheless, all groups have been able to find a tendency towards a lengthening of the survival time and the remission period after additional specific 'immunotherapy'. The most recent analysis of the study of Urbanitz et al. [438] showed that the survival data of the 45 patients, randomized to maintenance chemotherapy with or without specific immunotherapy, only reveal a trend in favor of those patients undergoing a combination of maintenance chemotherapy and specific immunotherapy. Until now, however, patients with more than 2 treatment courses of maintenance chemotherapy and immunotherapy already have a significant advantage concerning relapse-free survival (1020 days compared to 621 days) and remission duration (930 days compared to 409 days). If these positive results should persist in the course of the study, AML therapy with VCN-treated allogenous AML cells would be the first form of so-called active 'specific' tumor immunotherapy which has produced reproducible therapeutic successes of clinical relevance.

In accordance with animal experiments, there are also clinical indications that the therapeutic effectiveness of VCN-treated AML cells is directly dependent on the number of tumor cells administered. Thus, an injection of a total of only 5×10^8 VCN-treated AML cells distributed over 4 injection sites (instead of 1×10^{10} cells as in Bekesi et al. [34]) produced no significant differences to the relevant control group in respect to the remission period or the survival time [322].

Randomized prospective studies on specific 'immunotherapy' of solid tumors were carried out in melanoma patients by Simmons et al. [389] and Terry et al. [419], and in patients with carcinoma of the colon

Clinical Studies

by Gray (personal communication). In all the investigations, specific tumor therapy (VCN-treated tumor cells) was combined with non-specific immunotherapy (BCG). In none of these studies could the success of this combined immunotherapy be shown (see table XVII), possibly for similar reasons as those already discussed for a combination of specific and non-antigen specific immunotherapy of AML.

Combination of Tumor Cells and VCN

With regard to our experimental investigations, which showed the adjuvant effect of VCN on the one hand and the advantages of a combination of tumor cells and VCN, especially in chessboard vaccination, on the other, therapeutic investigations were carried out in leukemias (AML) and solid tumors (see table XVIII). In the case of AML, intradermal immunization with 10^9 allogenous AML cells mixed with 80 mU VCN and additionally applied to chemotherapy did nothing to change the results of chemotherapy alone (Rosenfeld, cited by [449]). There were no side-effects of the 'immunotherapy'.

Due to the lack of an adequate control group, the therapeutic effectiveness of chessboard vaccination with allogenous cells and VCN could not be established with any degree of certainty in the pilot study of AML [217]. There was, however, no great difference in comparison with historical controls. On the other hand, chessboard vaccination was tolerated without side-effects both locally and systematically.

Prospective randomized studies to establish the therapeutic effect of chessboard vaccination in AML have yet to be carried out. In view of the extremely high cell numbers (1×10^{10}) required monthly and per patient for successful specific immunotherapy with VCN-treated tumor cells in AML [34, 438], such a study is extremely difficult to perform but still an obvious necessity. With chessboard vaccination, always assuming the same degree of effectiveness, the required cell number could be diminished by a factor of 50 which would solve a large number of the logistical problems which at present limit specific tumor therapy with VCN-treated AML cells to a small number of tumor centers.

Investigations are at present being carried out in the specific tumor therapy of solid tumors with the aid of chessboard vaccination. They are mainly controlled randomized studies and concentrate on incurable diseases or those which are not affected by chemotherapy or can no

Table XVIII. Clinical studies on specific tumor therapy with combinations of tumor cells and VCN

Tumor (stage)	No. of pat.	Study protocol	Chemotherapy	Immunotherapy	Interim report	Results Final report	Authors
AML	29	random. 2 groups	Ara-C and ADM; VCR; Ara-C and TG	I) — II) 10^9 VCN blasts (allog.) and 80 mU VCN	no difference from control group		Rosenfeld et al. (EORTC, Paris) [364, 449]
AML	14	pilot study	Ara-C and ADM and TG/CP/CCNU	chessboard with VCN and allogenous cells		tendency towards improvement in comparison with control group	Lutz et al. (Wien) [217]
Stomach cancer – resectable	23	random. 2 groups	none	I) — II) chessboard with VCN and autolog. cells	started in 1981 I) relapse-free for 3 months, II) relapse-free for 8 months		Gürsel et al. (Istanbul) [364, 449]
– resectable and metastases		random. 2 groups	none	I) — II) chessboard with VCN and autolog. cells	started in 1981		Gürsel et al. (Istanbul) [364, 449]
Prostate gland carcinoma, pretreated	34	pilot study	varied (diethylstilbestrol; cyproteronacetate; bromocriptin)	Multiple chessboard with VCN and autolog. cells		significant drop of the prostate phosphatase and CEA; delay of tumor growth (?)	Rothauge et al. (Gießen) [130, 319]

Clinical Studies

Stage II	random. 2 groups		I) — II) chessboard (2×) with VCN and autolog. cells	started in 1981	Rothauge et al. (Gießen) [130, 319]
Colon carcinoma, Stage Duke B	60	random. 3 groups	5-FU; mitomyc.; Ara-C	I) — II) chemotherapy III) chessboard (3×) with allog. culture cells and VCN	started in 1977; up to now no difference between the groups (surviving 90%)	Rainer et al. (Wien) [282, 415]
Colon carcinoma, Stage Duke C	60	random. 3 groups	5-FU; mitomyc.; Ara-C	I) — II) chemotherapy III) chessboard (3×) with allog. culture cells and VCN	started in 1977; 40 months after surgery; significantly smaller mortality and smaller number of metastases after immunotherapy	Rainer et al. (Wien) [282]; Wunderlich et al. (Wien) [463]

longer be improved by it, such as carcinoma of the stomach, mammary carcinoma, and carcinoma of the prostate gland or colon.

Some of these investigations have already produced first promising results. Rothauge et al. [319] and Gutschank et al. [130] were able to show in an open, non-controlled study on more than 30 patients with uncontrollable, metastasizing carcinoma of the prostate gland that, after chessboard vaccination with autologous tumor cells and VCN, the serum levels of prostate phosphatase (determined radioimmunologically and enzymatically), acid phosphatase, and CEA diminished in comparison to their original values. Repetition of chessboard vaccination caused a further fall in the prostate phosphatase levels. In particular, in the case of CEA, this drop took place up to 7–10 weeks after chessboard vaccination and only after a temporary rise. This rise in the CEA level may be due to tumor necrosis as a result of chessboard vaccination. The fall in prostate phosphatase is generally considered to be a sign of tumor growth inhibition [415].

In comparison to a control group treated in the same center at the same time, a longer survival time could be found for patients vaccinated at least 3 times with the chessboard method [319]. These first clinical indications of the therapeutic effectiveness of chessboard vaccination are at present being tested in a randomized controlled study on prostate gland carcinoma in stage II [449].

In a further randomized prospective study, patients suffering from carcinoma of the colon, stages Duke B and C, were assigned to 3 groups. After resection of the tumor material, one group was not treated further; another group was given chemotherapy on 3 occasions consisting of arabinocytosine, 5-fluorouracil, and mitomycin C; and a third group was 'vaccinated' postoperatively 3 times within 3 months with a chessboard comprising inactivated cells from an allogenous colon carcinoma cell line and VCN [415]. More than 90% of the patients in stage Duke B are still alive. Differences between the treatment groups have not yet emerged. In contrast, for patients in stage C, after a 40-month observation period, the survival rate after 'immunotherapy' is approximately 72%, after chemotherapy approximately 51%, and in the untreated control group approximately 33% (see table XIX). Freedom from liver metastases is likewise different and is approximately 83% in the immunotherapy group, 81% in the chemotherapy group, and 54% in the control group. Moreover, the probability of being free of distant metastases was 80% in the immunotherapy group, 68% in the chemoimmunotherapy

Table XIX. Specific immunotherapy of colon carcinoma, stage Duke C, with chessboard vaccination

Treatment after tumor resection[a]	No. of patients	Surviving		Relapse rate	
		absolute	relative	absolute	relative
	22	7	32%	15	68%
Chemotherapy	19	10	53%	4	21%
Chessboard vaccination (allog. culture cells; VCN)	20	15	75%	6	30%

Results after an average observation period of 40 months
[a] Start of treatment within 6 weeks after radical operation. Treatment 3 times within 3 months
(Rainer et al. [282]; Wunderlich et al. [463])

group, and 52% in the control group [282, 463]. The strength of the skin reaction, which was also recorded in this study after chessboard vaccination, was dependent on the number of tumor cells, the amount of VCN admixed, the stage of the illness, and the number of vaccinations. Thus, there was no DTH reaction in any patient in stage C after the first chessboard vaccination, but there was a reaction after the second and third vaccination, and the local response was stronger after admixture of VCN within the chessboard [281].

All in all, these results clearly point to a therapeutic effectiveness of chessboard vaccination with allogenous culture cells and VCN for stage Duke C of colon carcinoma, whereby 'immunotherapy' is at least equally effective but much more tolerable than chemotherapy. Thus, unlike chemotherapy, after which 37% of the patients complained of slight and severe disturbances in their general well-being, there were no side-effects in the patients undergoing 'immunotherapy'. Thus, it has been concluded [463] that, whatever the results of the trial in patients with colon carcinoma will be after a longer period of observation, the prolongation of the disease-free interval will remain as a valuable contribution of adjuvant therapy towards a better quality of life in patients who have been operated on for an advanced stage of colorectal cancer.

The therapeutic results with chessboard vaccination so far, especially in carcinoma of the colon, are so evident and distinct that we must certainly confirm them by means of further and, if necessary, more extensive studies. In view of the disappointing clinical experience in tumor immunotherapy with non-specific immunomodulators, specific tumor therapy in the form of chessboard vaccination with VCN can only be recommended as a form of therapy for any tumor illness after such confirmation has been found. It would appear feasible to repeat the studies since there have been neither indications of side-effects nor of an acceleration of tumor growth after chessboard vaccination.

Summary

The cell physiological, immunological, experimental, and clinical data of specific tumor therapy with the use of Vibrio cholerae neuraminidase are reviewed.

Neuraminidase induces obvious biological effects caused by its enzymatic activity to cleave membrane-bound N-acetyl neuraminic acid:

— VCN leads to an enhancement of cell-to-cell contact by a decrease in the repelling negative cell charge. VCN also exposes receptors on the cell membrane, for instance for the Fc-fragment of IgG, for LAF and for erythropoietin.

— VCN does not increase the exposure of cell membrane antigens, measured by the binding or absorption of specific antibodies. VCN demasks crypt-antigens such as the Thomsen-Friedenreich antigen, not only on erythrocytes but also on tumor cells of various species. However, there are no indications that this antigen might be of decisive importance in tumor therapy with VCN.

— VCN treatment of cells, including tumor cells, can enhance their immunogenicity in vitro as well as in vivo. The application of VCN-treated tumor cells may induce an antigen-specific regression of tumors. However, this effect is difficult to reproduce. Moreover, depending on the cell dose applied, enhancement of tumor growth may be induced.

— VCN sticks to the cell membrane. Inhibition experiments make it probable that this adherence to the cell membrane does not appear to take place via the enzymatically active part of the VCN molecule and that terminal galactose is a decisive component of the membrane structure to which VCN binds.

— VCN on the cell membrane may present an exogenous antigen which is able to decisively alter cell antigenicity. Membrane-bound VCN can also have an adjuvant activity on the immune reaction against cell-membrane antigens as VCN exhibits adjuvant activity for both cells and neuraminic acid free antigens. Adjuvant activity is dependent on the amount of antigen (number of cells) and on the enzymatic activity and amount of VCN admixed to the antigen.

— Lymphocytes (T-cells), but not macrophages or granulocytes, were identified as target cells for adjuvant activity. VCN probably increases the reactivity of lymphocytes for immune mediators.

On the basis of these results and in special consideration of the adjuvant effect of neuraminidase, a new application scheme called chessboard vaccination was developed for tumor therapy with VCN. Chessboard vaccination consisted in the simultaneous intradermal injection of increasing numbers of tumor cells, each mixed with increasing amounts of VCN. Chessboard vaccination revealed to be at least equally effective (mammary tumor in dogs) or more effective (Lewis lung adenocarcinoma) in tumor therapy than tumor cells conventionally treated with VCN. No indications of an acceleration of tumor growth could be found. The reasons for the improved therapeutic effect of chessboard vaccination in comparison to VCN-treated tumor cells and the consequent smaller risk of treatment-induced enhancement of tumor growth are unclear.

It is speculated that the therapeutically effective cell doses may dominate over mixtures with immunosuppressing effects in case a combination of possibly stimulating and suppressing cell doses is applied in the course of chessboard vaccination. Moreover, the individual reactive situation of the organism may be better taken into account with chessboard vaccination than by the application of one certain amount of cells.

Clinical investigations have now given indications that, in selected human tumors, chessboard vaccination has a relevant therapeutical effect. It remains to be seen whether this effect is reproducible and whether we, thus, have the possibility of specific tumor immune therapy for humans.

Acknowledgments

I am indebted to Prof. V. Schirrmacher, Institute of Tumor Immunology, German Cancer Research Center, Heidelberg; Prof. R. Kurth, Paul Ehrlich Institute, Frankfurt; and Prof. W. Müller-Ruchholtz, Institute for Immunobiology, University of Kiel, for critical reading of the manuscript.

The manuscript has been typewritten and corrected by Mrs. S. Lehnert, H. Donges, and I. Gimpel. Their excellent help, which proved to be essential for the final outcome, is gratefully acknowledged. Moreover, I thank Dr. J. Henning for her help in the bibliography.

References

1 Abandowitz, H. M.: Neuraminidase effect on the growth of a transplantable nickel sulfide induced rat tumor. J. Jap. Med. Sci. Biol. *31:* 421–424 (1978).
2 Abo, T.; Yamaguchi, T.; Shimizu, F.; Kumagai, K.: Studies of surface immunoglobulins on human B lymphocytes. II. Characterization of a population of lymphocytes lacking surface immunoglobulins but carrying Fc receptors (SIg⁻Fc⁺ cell). J. Immunol. *117:* 1781–1787 (1976).
3 Ada, G. L.; Stone, J. D.: Electrophoretic studies of virus red cell interaction: Mobility gradient of cells treated with viruses of the influenza group and the receptor destroying enzyme of V. cholerae. Br. J. Exp. Path. *31:* 263–274 (1950).
4 Ada, G. L.; French, E. L.; Purification of bacterial neuraminidase (receptor-destroying enzyme). Nature *183:* 1740–1741 (1959).
5 Adler, W. H.; Osunkoya, B. G.; Takiguchi, T.; Smith, R. T.: The interactions of mitogens with lymphoid cells and the effect of neuraminidase on the cells responsiveness to stimulations. Cell. Immunol. *3:* 590–605 (1972).
6 Akiyoshi, T.; Kawaguchi, M.; Miyazaki, S.; Tsuji, H.: Lymphocyte blastogenic response to autologous tumor cells pretreated with neuraminidase in patients with gastric carcinoma. Oncology *37:* 309–313 (1980).
7 Albright, L.; Madigan, J. C.; Gaston, M. R.; Houchens, D. P.: Therapy in an intracerebral murine glioma model, using bacillus Calmette-Guérin, neuraminidase-treated tumor cells, and 1-(2-chloroethyl)-3-cyclohexyl-1-nitrosourea. Cancer Res. *35:* 658–665 (1975).
8 Allen, J. M.; Cook, G. M.: A study of the attachment phase of phagocytosis by murine macrophages. Exp. Cell Res. *59:* 105–116 (1970).
9 Alley, C. D.; Snodgrass, M. J.: Effectiveness of neuraminidase in experimental immunotherapy of two murine pulmonary carcinomas. Cancer Res. *37:* 95–101 (1977).
10 Alley, C. D.; Snodgrass, M. J.: Effect of inoculation with neuraminidase-treated tumor cells on macrophage cytotoxicity in vitro. Cancer Res. *38:* 2332–2338 (1978).
11 Altevogt, P.; Fogel, M.; Cheingsong-Popov, R.; Dennis, J.; Robinson, P.; Schirrmacher, V.: Different patterns of lectin binding and cell surface sialylation detected on related high- and low-metastatic tumor lines. Cancer Res. *43:* 5138–5144 (1983).
12 Ambrose, E. J.; James, A. M.; Lowice, J. H.: Differences between the electrical charge carried by normal and homologous tumor cells. Nature, Lond. *177:* 576–577 (1956).
13 Aminoff, D.; Von der Bruegge, W. F.; Bell, W. C.; Sarpolis, K.; Williams, R.: Role of sialic acid in survival of erythrocytes in the circulation: Interaction of

References

neuraminidase-treated and untreated erythrocytes with spleen and liver at the cellular level. Proc. natn. Acad. Sci. USA *74:* 1521–1524 (1977).

14 Anderson, A. C.: Parameters of mammary gland tumors in aging beagles. J. Am. vet. med. Ass. *147:* 1653–1654 (1965).

15 Anglin, J. H., Jr.; Lerner, M. P.; Nordquist, R. E.: Blood group-like activity released by human mammary carcinoma cells in culture. Nature, Lond. *269:* 254–255 (1977).

16 Apffel, C. A.; Arnason, B. G.; Peters, J. H.: Induction of tumour immunity with tumour cells treated with iodoacetate. Nature, Lond. *209:* 694–696 (1966).

17 Arala-Chaves, M. P.; Hope, L.; Korn, J. H.; Fudenberg, H.: Role of adherent cells in immune responses to phytohemagglutinin and concanavalin A. Eur. J. Immunol. *8:* 77–81 (1978).

18 Ascher, M. S.; Parker, D.; Turk, J. L.: Modulation of delayed-type hypersensitivity and cellular immunity of microbial vaccines: Effects of cyclophosphamide on the immune response to tularemia vaccine. Infect. Immun. *18:* 318–323 (1977).

19 Ashwell, G.; Morell, A. G.: The role of surface carbohydrates in the hepatic recognition and transport of circulating glycoproteins. Adv. Enzymol. *41:* 99–128 (1974).

20 Babior, B. M.; Kipnes, R. S.; Curnatte, J. T.: Biological defense mechanisms: the production of leukocytes of superoxide, a potential bactericidal agent. J. Clin. Invest. *52:* 741–744 (1973).

21 Bach, J. F.; Dormont, J.; Dardenne, M.; Balner, H.: In vitro rosette-inhibition by antihuman antilymphocyte serum. Transplantation *8:* 265–280 (1969).

22 Bagshawe, K. D.; Currie, G. A.: Immunogenicity of L1210 murine leukemia cells after treatment with neuraminidase. Nature, Lond. *218:* 1254–1255 (1968).

23 Balke, E.; Scharmann, W.; Drzeniek, R.: Die Bestimmung des Molekulargewichtes bakterieller Neuraminidasen mit Hilfe der Gel-Filtration. Zbl. Bakt. Hyg., I. Abt. Orig. A *229:* 55–67 (1974).

24 Barinsky, I. F.; Kobrinsky, G. D.: An inhibitory action of neuraminidase of cholerae vibrio in mouse Rauscher leukemia. Bull. Eksp. Biol. Med. *82:* 1357–1359 (1976).

25 Barth, R. F.; Singla, O.: Alterations in the immunogenicity and antigenicity of mammalian erytrhocytes following treatment with neuraminidase (3770). Proc. Soc. exp. Biol. Med. *145:* 168–172 (1974).

26 Barton, N. W.; Rosenberg, A.: Action of Vibrio cholerae neuraminidase (sialidase) upon the surface of intact cells and their isolated sialolipid components. J. Biol. Chem. *248:* 7353–7358 (1973).

27 Baxley, G.; Bishop, G. B.; Cooper, A. G.; Wortis, H. H.: Rosetting of human red blood cells to thymocytes and thymus-derived cells. Clin. exp. Immunol. *15:* 385–392 (1973).

28 Bekesi, J. G.; St.-Arneault, G.; Holland, J. F.: Increase of leukemia L-1210 immunogenicity by Vibrio cholerae neuraminidase treatment. Cancer Res. *31:* 2130–2132 (1971).

29 Bekesi, J. G.; Roboz, J. P.; Walter, L.; Holland, J. F.: Stimulation of specific immunity against cancer by neuraminidase-treated tumor cells. Behring Inst. Mitt. *55:* 309–321 (1974).

30 Bekesi, J. G.; Holland, J. F.: Chemoimmunotherapy of leukemia in man and experimental animals. Int. Conf. on Immunotherapy of Cancer, New York, Nov. 5–7 (1975).

31 Bekesi, J. G.; Roboz, J. P.; Holland, J. F.: Therapeutic effectiveness of neuraminidase-treated tumor cells as an immunogen in man and experimental animals with leukemia. Ann. N. Y. Acad. Sci. *277:* 313–331 (1976).
32 Bekesi, J. G.; Holland, J. F.: Active immunotherapy in leukaemia with neuraminidase-modified leukaemic cells. Recent Results Cancer Res. *62:* 78–89 (1977).
33 Bekesi, J. G.; Holland, J. F.: Immunotherapy of acute myelocytic leukaemia with neuraminidase-treated myeloblast and MER; in Rainer, Immunotherapy of malignant diseases, pp. 375–379 (Schattauer, Stuttgart 1978).
34 Bekesi, J. G.; Holland, J. F.: Immunotherapy of acute myelocytic leukaemia with neuraminidase-treated myeloblasts in Acute leukemias: Prognostic factors and treatment strategies (Abstract No. 13). Int. Symp., Münster, Febr. 23–25 (1986).
35 Bentwich, Z.; Douglas, S. D.; Skutelsky, E.; Kunkel, H. G.: Sheep red cell binding to human lymphocytes treated with neuraminidase: Enhancement of T cell binding and identification of a subpopulation of B cells. J. exp. Med. *137:* 1532–1537 (1973).
36 Berger-Du Bois, F.; Des Gouttes, D.; Isliker, H.: Effect de la neuraminidase sur certaines propriétés biologiques des immunoglobulines G. Ann. Inst. Pasteur *119:* 8–16 (1970).
37 Beyer, C. F.; Bowers, W. E.: Lymphocyte transformation induced by chemical modification of membrane components. II. Effect of neuraminidase treatment of responder cells on proliferation and cytotoxicity in indirect stimulation. J. Immunol. *121:* 1790–1798 (1978).
38 Binder, P.; Perrot, L.; Beaudry, Y.; Bottex, C.; Fontanges, R.: Cancérologie. – Étude, chez le rat, des effets de la neuraminidase, bactérienne et virale, en injection intratumorale. C. R. Acad. Sci. Paris *281:* 1545–1547 (1975).
39 Biozzi, G.; Stiffel, C.; Mouton, D.; Liacopoulos-Briot, M.; Decreusefond, C.; Bouthillier, Y.: Étude du phénomène de l'immuno-cyto-adhérence au cours de l'immunisation. Ann. Inst. Pasteur *110:* 7–32 (1966).
40 Bird, G. W. G.; Anti-T in peanuts. Vox Sang. *9:* 748–749 (1964).
41 Blix, F. G.; Gottschalk, A.; Klenk, E.: Proposed nomenclature in the field of neuraminidase and sialic acids. Nature *179:* 1088–1089 (1957).
42 Bliznakov, E. G.: Immunostimulation or immunodepression? Biomedicine *26:* 73–76 (1977).
43 Bolhuis, R. L. H.; Schuit, H. R. E.; Nooyen, A. M.; Ronteltap, C. P. M.: Characterization of natural killer (NK) cells and killer (K) cells in human blood: discrimination between NK and K cell activities. Eur. J. Immunol. *8:* 731–740 (1978).
44 Boschmann, T. A. C.; Jacobs, J.: The influence of ethylenediaminetretraacetate on various neuraminidase. Biochem. Z. *342:* 532–541 (1965).
45 Bosslet, K.; Lüben, G.; Stark, M.; Sedlacek, H. H.: Characterization of an individual specific small cell lung carcinoma-associated antigen. Behring Inst. Mitt. *78* 133–138 (1985).
46 Brandt, A. E.; Jameson, A. K.; Pincus, J. H.: Characterization and use of neuraminidase-modified L1210 plasma membranes for protection against tumor growth. Cancer Res. *41:* 3077–3081 (1981).
47 Braun, D. P.; Harris, J. E.: Modulation of the immune response by chemotherapy. Pharmacol. Ther. *14:* 89–122 (1981).
48 Bray, J.; Lemieux, R. U.; McPherson, T. A.: Use of a synthetic hapten in the demon-

stration of the Thomsen-Friedenreich (T) antigen on neuraminidase-treated human red blood cells and lymphocytes. J. Immunol. *126:* 1966–1969 (1981).
49 Brazil, J.; McLaughlin, H.: The modification of the growth characteristics of the Landschütz ascites tumour by neuraminidase. Eur. J. Cancer *14:* 757–760 (1978).
50 Bretscher, M. S.; Raff, M. C.: Mammalian plasma membranes. Nature *258:* 43–49 (1975).
51 Brinckerhoff, C. E.; Lubin, M.: Increased tumor immunity in mice inoculated with myconomycin A-treated B16-melanoma cells. Cancer Res. *38:* 3668–3673 (1978).
52 Burger, M. M.: Surface changes in transformed cells detected by lectins. Fed. Proc. *32:* 91–101 (1973).
53 Celik, C.; Lewis, D. A.; Goldrosen, M. H.: Demonstration of immunogenicity with the poorly immunogenic B16 melanoma. Cancer Res. *43:* 3507–3510 (1983).
54 Cihak, J.; Ziegler, H. W.; Kölsch, E.: Regulation of immune responses against the syngeneic ADJ-PC-5 plasmacytoma in Balb/c mice. II. Suppression of T-cell cytotoxicity by pretreatment of mice with subimmunogenic doses of tumour cells. Immunology *43:* 145–152 (1981).
55 Codington, J. F.; Sanford, B. H.; Jeanloz, R. W.: Glycoprotein coat of the TA3 cell. I. Removal of carbohydrate and protein material from viable cells. J. natn. Cancer Inst. *45:* 637–647 (1970).
56 Cohnen, G.; Fischer, K.; Brittinger, G.: Human T-lymphocyte rosette formation: Inhibition by cytochalasin B. Immunology *29:* 337–341 (1975).
57 Colman, P. M.; Ward, C. W.: Structure and diversity of influenza virus neuraminidase. Curr. Top. Microbiol. Immunol. *114:* 177–255 (1985).
58 Constantopoulos, A.: The distribution of sialic acid on human erythrocyte membrane. Cytobios *7:* 97–102 (1973).
59 Cook, G. M.; Heard, D. H.; Seaman, G. V.: Sialic acids and the electrokinetic charge of the human erythrocyte. Nature *191:* 44–47 (1961).
60 Cook, G. M.; Jacobsen, W.: The electrophoretic mobility of normal and leukaemic cells of mice. Biochem. J. *107:* 549–557 (1968).
61 Cormack, D.: Effect of enzymatic removal of cell surface sialic acid on the adherence of Walker 256 tumor cells to mesotheial membrane. Cancer Res. *30:* 1459–1466 (1970).
62 Crumpton, M. J.; Davies, D. A. C.; Hutchinson, A. M.: The serological specificities of Pasteurella pseudotuberculosis somatic antigens. J. Gen. Microbiol. *18:* 129–139 (1958).
63 Cuatrecasas, P.; Illiano, G.: Membrane sialic acid and the mechanism of insulin action in adipose tissue cells. Effects of digestion with neuraminidase. J. Biol. Chem. *246:* 4938–4946 (1971).
64 Cullen, S. E.; David, C. S.; Shreffler, D. C.; Natheson, S. G.: Membrane molecules determined by the H-2-associated immune response region: isolation and some properties. Proc. natn. Acad. Sci. USA *71:* 648–652 (1974).
65 Cullen, S. E.; Freed, J. H.; Nathenson, S. G.: Structural and serological properties of murine Ia alloantigens. Transplant. Rev. *30:* 236–270 (1976).
66 Currie, G. A.: Masking of antigens on the Landschütz ascites tumor. Lancet *ii:* 1336–1338 (1967).
67 Currie, G. A.; Bagshawe, K. D.: The effect of neuraminidase on the immunogenicity

of the Landschütz ascites tumor. Site and mode of action. Br. J. Cancer 22: 588–594 (1968).
68 Currie, G. A.; Bagshawe, K. D.: The role of sialic acid in antigenic expression: Further studies of the Landschütz ascites tumor. Br. J. Cancer 22: 843–853 (1968).
69 Currie, G. A.; Bagshawe, K. D.: Tumor-specific immunogenicity of methylcholanthrene-induced sarcoma cells after incubation in neuraminidase. Br. J. Cancer 23: 141–149 (1969).
70 Dalmasso, A. P.; Müller-Eberhard, H. J.: Interaction of autologous complement with red cells in the absence of antibody. Proc. Soc. exp. Biol. Med. 117: 643–650 (1964).
71 D'Arville, C. N.; Pierrepoint, C. G.: The demonstration of oestrogen, androgen and progestagen receptors in the cytosol fraction of canine mammary tumours. Eur. J. Cancer 15: 875–883 (1979).
72 Davies, J. W.; Yue, K. T. N.; Phillips, P. E.: The effect of neuraminidase on platelet aggregation induced by ADP, norepinephrine, collagen or serotonin. Thromb. Diathes. Haemorrh. 28: 221–227 (1972).
73 Defendi, V.; Gasic, G.: Surface mucopolysaccharides of polyoma virus transformed cells. J. Cell. Comp. Physiol. 62: 23–31 (1963).
74 Deman, J. J.; Bruyneel, E. A.; Mareel, M. M.: A study on the mechanism of intercellular adhesion. Effects of neuraminidase, calcium and trypsin on the aggregation of suspended HeLa cells. J. Cell Biology 60: 641–652 (1974).
75 Dennis, J. W.; Donaghue, T. P.; Kerbel, R. S.: An examination of tumor antigen loss in spontaneous metastases. Invasion Metastasis 1: 111–125 (1980).
76 Dennis, J. W.; Kerbel, R. S.: Characterization of a deficiency in fucose metabolism in lectin-resistant variants of a murine tumor showing altered tumorigenic and metastatic capacities in vivo. Cancer Res. 41: 98–104 (1981).
77 Dennis, J.; Waller, C.; Timpl, R.; Schirrmacher, V.: Surface sialic acid reduces attachment of metastatic tumour cells to collagen type IV and fibronectin. Nature, Lond. 300: 274–276 (1982).
78 Despont, J. P.; Abel, C. A.; Grey, H. M.: Sialic acid and sialyltransferases in murine lymphoid cells: Indicators of T cell maturation. Cell. Immunol. 17: 487–494 (1975).
79 Dewitt, C. W.; Zell, E. A.: Sialic acids (N,7-0-diacetyl-neuraminic acid) in Escherichia coli. II. Their presence on the cell surface and relationship to k-antigen. J. Bact. 82: 849–851 (1961).
80 Diamantstein, T.; Willinger, E.; Reiman, J.: T-suppressor cells sensitive to cyclophosphamide and to its in vitro active derivative 4-hydroperoxycyclophosphamide control the mitogenic response of murine splenic B cells to dextran sulfate. J. exp. Med. 150: 1571–1576 (1979).
81 Dolen, J. G.; Han, T.: Enhancement of human normal and leukemic B lymphocyte rosette formation with neuraminidase-treated mouse red blood cells. IRCS Med. Science 5: 594 (1977).
82 Doré, J. F.; Hadjiyannakis, M. J.; Guibout, C.; Coudert, A.; Marholev, L.; Imai, K.: Use of enzyme-treated cells in immunotherapy of a murine leukemia. Lancet viii: 600–601 (1973).
83 Dorn, C. R.; Taylor, D. O. N.; Frye, F. L.; Hibbard, H. H.: Survey of animal neoplasms in Alameda and contra costa counties, California. I. Methodology and description of cases. J. natn. Cancer Inst. 40: 295–305 (1968).

References

84 Dorrington, K. J.: Properties of the Fc-receptor on macrophages and monocytes. Immunol. Commun. 5: 263–280 (1976).
85 Drzeniek, R.: Differences in splitting capacity of virus and Vibrio cholerae neuraminidases on sialic acid type substrates. Biochem. Biophys. Res. Commun. 26: 631–638 (1967).
86 Drzeniek, R.: Viral and bacterial neuraminidases. Curr. Top. Microbiol. Immunol. 59: 35–74 (1972).
87 Drzeniek, R.: Substrate specificity of neuraminidases. Histochem. J. 5: 271–290 (1973).
88 Du Bois, M. J. G.; Huismans, D. R.; Schellckens, P. T. A.; Eijsvoogel, V. P.; Investigation and standardization of the conditions for microlymphocyte cultures. Tissue Antigens 3: 402–409 (1973).
89 Egeberg, J.; Jensen, O. A.: The effect of neuraminidase-treated tumor cells on the growth of transplantable malignant melanoma of the Syrian golden hamster (mesocricetus auratus). Int. Rev. Cytol. 2 (suppl.): 1573 (1974).
90 Eisenberg, S.; Ben-Or, S.; Dolanski, F.: Electrokinetic properties of cells in growth processes. I. The electronic behavior of liver cells during regeneration and postnatal growth. Exp. Res. 26: 451–461 (1962).
91 Faraci, R. P.: In vitro demonstration of altered antigenicity of metastases from a primary methylcholanthrene-induced sarcoma. Surgery 76: 469–473 (1974).
92 Faraci, R. P.; Marrone, A. C.; Ketham, A. S.: Anti-tumor immune response following infection of neuraminidase-treated sarcoma cells. Ann. Surgery 181: 359–362 (1975).
93 Feldman, D. G.; Gross, L.: Electron microscopic study of spontaneous mammary carcinomas in cats and dogs: virus-like particles in cat mammary carcinomas. Cancer Res. 31: 1261–1267 (1971).
94 Ferguson, R.; Anderson, S. M.; Schmidtke, J. R.; Simmons, R. L.: Effect of Vibrio cholerae neuraminidase on the generation of cell-mediated cytotoxicity in vitro. J. Immunol. 117: 2150–2157 (1976).
95 Fidler, I. J.: In vitro studies of cellular-mediated immunostimulation of tumor growth. J. natn. Cancer Inst. 50: 1307–1312 (1973).
96 Fidler, I. J.; Brodey, R. S.: A necroscopy study of canine malignant mammary neoplasms. J. Am. vet. med. Ass. 151: 710–715 (1976).
97 Fidler, I. J.; Kahn, J. M.; Montgomery, P. C.: Effect of neuraminidase on the rat: one-way mixed lymphocyte interaction. Immunol. Commun. 2: 573–583 (1973).
98 Fischer, K.; Poschmann, A.; Stegner, H. E.: Nachweis von Mammarkarzinom-Zellen mit der Immunfluoreszenztechnik. Dt. med. Wschr. 102: 1227 (1977).
99 Flowers, H. M.; Sharon, N.; Glycosidase properties and application to the study of complex carbohydrates and cell surfaces. Adv. Enzymol. 48: 29–95 (1979).
100 Flye, M. W.; Reisner, E. G.; Amos, D. B.: The in vitro effect of neuraminidase on human lymphocytes. J. Surg. Res. 15 (2): 96–99 (1973).
101 Fogel, M.; Gorelik, E.; Segal, S.; Feldman, M.: Differences in cell surface antigen of tumor metastases and those of the local growth. J. natn. Cancer Inst. 62: 585–588 (1979).
102 Fogel, M.; Altevogt, P.; Schirrmacher, V.: Metastatic potential severely altered by changes in tumor cell adhesiveness and cell-surface sialylation. J. exp. Med. 157: 371–376 (1983).

103 Fraser, K. B.: The information of antibody. A study of the relationship between a normal and an immune haemagglutinin. J. Pathol. Bact. *LXX:* 13–33 (1955).
104 Friedenreich, V.: Untersuchungen über das von O. Thomsen beschriebene vermehrungsfähige Agens als Veränderer des isoagglutinatorischen Verhaltens der roten Blutkörperchen. Z. Immunitätsforsch. *55:* 84–101 (1928).
105 Friedenreich, V.: Die serologische Auffassung des Thomsenschen Blutkörperchenrezeptors. Acta pathol. microbiol. scand. (Suppl.) *V:* 68 (1930).
106 Fröland, S. S.: Binding of sheep erythrocytes to human lymphocytes. A probable marker of T lymphocytes. Scand. J. Immunol. *1:* 269–280 (1972).
107 Froese, G.; Berczi, I.; Sehon, A. H.: Brief communication: Neuraminidase-induced enhancement of tumor growth in mice. J. natn. Cancer Inst. *52:* 1905–1908 (1974).
108 Fuhrmann, G. F.: Cytopherograms of normal proliferating and malignant rat liver cells; in Ambrose, Cell electrophoresis, p. 92 (Churchill, London 1965).
109 Fuhrmann, G. F.; Granzer, E.; Kübler, W.; Rueff, F.; Ruhenstroth-Bauer, G.: Neuroaminosäurenbedingte Strukturunterschiede der Zellmembranen normaler und maligner Leberzellen. Z. Naturf. *17b:* 610–613 (1962).
110 Galili, U.; Schlesinger, M.: The formation of stable E rosettes after neuraminidase treatment of either human peripheral blood lymphocytes or of sheep red blood cells. J. Immunol. *112:* 1628–1634 (1974).
111 Galili, U.; Schlesinger, M.: Regulation of the cytotoxic effect of human 'normal killer cells' on tumor cell lines by neuraminidase-treated T-lymphocytes. Cancer Immunol. Immunother. *4:* 33–39 (1978).
112 Gautam, S.; Aikat, B. K.: Immunotherapy of methylcholanthrene-induced and spontaneous tumours in mice by use of tumour vaccine, neuraminidase and BCG. Indian J. med. Res. *64 (3):* 472–481 (1976).
113 Ghose, T.; Guclu, A.; Tai, J.; Mammen, M.; Norvell, S. T.: Immunoprophylaxis and immunotherapy of EL4 lymphoma. Eur. J. Cancer *13:* 925–935 (1977).
114 Gielen, W.: Neuraminidase in higher organisms. Behring Inst. Mitt. *55:* 85–88 (1974).
115 Gilbertsen, R. B.; Metzgar, R. S.: Human T and B lymphocytes rosette tests: Effect of enzymatic modification of sheep erythrocytes (E) and the specificity of neuraminidase-treated E. Cell. Immunol. *24:* 97–108 (1976).
116 Gill, H. K.; Liew, F. Y.: Regulation of delayed-type hypersensitivity. III. Effect of cyclophosphamide on the suppressor cells for delayed-type hypersensitivity to sheep erythrocytes in mice. Eur. J. Immunol. *8:* 172–176 (1978).
117 Girard, J. P.; Fernandes, B.; Studies on the mitogenic activity of trypsin, pronase and neuraminidase on human peripheral blood lymphocytes. Eur. J. clin. Invest. *6:* 347–353 (1976).
118 Glaser, M.: Regulation of specific cell-mediated cytotoxic response against SV40-induced tumor-associated antigens by depletion of suppressor T cells with cyclophosphamide in mice. J. exp. Med. *149:* 774–779 (1979).
119 Glick, M. C.: Isolation and characterization of surface membrane glycoproteins from mammalian cells. Methods Membrane Biol. *2:* 157–204 (1974).
120 Glöckner, M. W.; Newman, R. A.; Dahr, W.; Uhlenbruck, G.: Alkali-labile oligosaccharides from glycoproteins of different erythrocyte and milk fat globule membranes. Biochim. Biophys. Acta *443:* 402–413 (1976).
121 Glöckner, W. M.; Kaulen, H. D.; Uhlenbruck, G.: Immunochemical detection of the

Thomsen-Friedenreich antigen (T-antigen) on platelet plasma membranes. Thromb. Haemostas. *39:* 186–192 (1978).
122 Görög, P. et al.: Increased adhesiveness of granulocytes in rabbit ear chamber blood vessels perfused with neuraminidase. Microvasc. Res. *23:* 380–384 (1982).
123 Gordon, S.; Todd, J.; Cohn, Z. A.: In vitro synthesis and secretion of lysozyme by mononuclear phagocytes. J. exp. Med. *139:* 1228–1248 (1974).
124 Gottschalk, A.: Neuraminidase: The specific enzyme of influenza virus and Vibrio cholerae. Biochim. Biophys. Acta *23:* 645–646 (1975).
125 Gottschalk, A.: Neuraminidase: Its substrate and mode of action. Adv. Enzymol. Interscience *20:* 135–146 (1958).
126 Gottschalk, A.; Bhargava, A. S.: Neuraminidase; in Bayer, The enzymes, pp. 321–342 (Academic Press, New York 1971).
127 Greenberg, J.; Packham, M. A.; Cazenave, J.-P.; Reimers, H.-J.; Mustard, J. F.: Effects on platelet function of removal of platelet sialic acid by neuraminidase. Lab. Invest. *32:* 476–484 (1975).
128 Grimes, W. J.: Sialic acid transferases and sialic acid levels in normal and transformed cells. Biochemistry *9:* 5083–5092 (1970).
129 Grothaus, E. A.; Flye, M. W.; Yunis, E.; Amos, B. D.: Human lymphocyte antigen reactivity modified by neuraminidase. Science *173:* 542–544 (1971).
130 Gutschank, S.; Torhauge, C. F.; Kraushaar, J.; Gutschank, W.; Sedlacek, H. H.: Das entgleiste metastasierende Prostatakarzinom. Münch. med. Wschr. *123:* 133–134 (1981).
131 Haegert, D. G.: Phagocytic peripheral blood monocytes from rabbits and humans express membrane receptors specific for IgM molecules: evidence that incubation with neuraminidase exposes cryptic IgM (Fc) receptors. Clin. exp. Immunol. *35:* 484–490 (1979).
132 Haegert, D. G.: Demonstration of surface membrane immunoglobulin on L-lymphocytes by the mixed antiglobulin rosetting reaction (MARR) and by the direct antiglobulin rosetting reaction (DARR). Immunology *38:* 459–465 (1979).
133 Hagmayer, G.: Therapie der spontanen Mammatumoren des Hundes mit autologen Zellen und Neuraminidase. Inaugural-Diss., Giessen (1982).
134 Hakomori, S. I.: Glycolipid changes associated with malignant transformation; in Wallach, Hoelzl, Fischer, The dynamic structure of cell membranes. 22nd Colloquium der Ges. für Biologische Chemie, Mosbach/Baden 1971, pp. 65–96 (Springer, New York 1971).
135 Hakomori, S. I.; Murakami, W. T.: Glycolipids of hamster fibroblasts and derived malignant transformed cell lines. Proc. natn. Acad. Sci. USA *59:* 254–261 (1968).
136 Hakomori, S.; Kannagi, R.: Glycosphingolipids as tumor-associated and differentiation markers. J. natn. Cancer. Inst. *71:* 231–251 (1983).
137 Han, T.: Enhancement of mixed lymphocyte reactivity by neuraminidase. Transplantation *14:* 515–517 (1972).
138 Han, T.: Enhancement of in vitro lymphocyte response by neuraminidase. Clin. exp. Immunol. *13:* 165–170 (1973).
139 Han, T.: Enhancement of delayed skin hypersensitivity by neuraminidase in cancer patients. Clin. exp. Immunol. *18:* 95–100 (1974).
140 Han, T.: Specific effect of neuraminidase on blastogenic response of sensitized lymphocytes. Immunology *28:* 283–286 (1975).

141 Han, T.: Neuraminidase-mediated enhancement of lymphocyte response to pokeweed mitogen: exclusive action of neuraminidase on responding T lymphocytes. IRCS Medical Science 6: 152 (1978).
142 Han, T.; Dadey, B.: Isolation of two functionally distinct subpopulations of human peripheral blood T lymphocytes by sheep and human erythrocyte rosetting techniques. IRCS Medical Science 6: 93 (1978).
143 Hanfland, P.; Uhlenbruck, G.: Einige Aspekte zur Biochemie der Tumorzelle. Internist 18: 269–276 (1977).
144 Harthus, H. P.; Johannsen, R.; Ax, W.: Lymphocyte sensitization in tumor-bearing rats: EMT versus MLTC. Immunobiology 156: 268–269 (1979).
145 Harthus, H. P.; Ax, W.: Electrophoretic mobility test (EMT): Studies on lymphocyte response and mechanism of the test using a rat tumor model. Immunobiology 158: 151–172 (1981).
146 Harvarth, L.; Amirault, H. J.; Anderson, B. R.: Chemiluminescence of human and canine polymorphonuclear leukocytes in the absence of phagocytosis. J. clin. Invest. 61: 1145–1154 (1978).
147 Hauschka, T. S.; Weiss, L.; Holdridge, B. A.: Karyotypic and surface features of murine TA3 carcinoma cells during immunoselection in mice and rats. J. natn. Cancer Inst. 47: 343–359 (1971).
148 Heddle, R. J.; Knop, J.; Steele, E. J.; Rowley, D.: The effect of lysozyme on the complement-dependent bactericidal action of different antibody classes. Immunology 28: 1061 (1975).
149 Henson, P. M.: Interaction of cells with immune complexes: Adherence release of constituents and tissue injury. J. exp. Med. 134: 114 (1971).
150 Henson, P. M.: Complement-dependent platelet and polymorphonuclear leukocyte reactions. Transplant. Proc. 6: 27–31 (1974).
151 Higgins, T. J.; Parish, C. R.: Extraction of the carbohydrate-defined class of Ia antigens from murine spleen cells and serum. Mol. Immunol. 17: 1065–1073 (1980).
152 Hoessli, D.; Bron, C.; Pink, J. R. L.: T-lymphocyte differentiation is accompanied by increase in sialic acid content of Thy-1 antigen. Nature 283: 576–578 (1980).
153 Holland, J. F.; St. Arneault, G.; Bekesi, G.: Combined chemo- and immunotherapy of transplantable and spontaneous murine leukemia. Proc. Am. Ass. Cancer Res. 13: 83 (1972).
154 Holland, J. F.; Bekesi, J. G.: Immunotherapy of human leukemia with neuraminidase-modified cells. Med. Clins. N. Am. 60: 539–549 (1976).
155 Holland, J. F.; Bekcsi, J. G.; Guttner, J.; Glidewell, O.: Chemoimmunotherapy in acute myelocytic leukemia. Israel J. med. Sci. 13: 694–700 (1977).
156 Holland, J. F.; Bekesi, J. G.: Comparison of chemotherapy with chemotherapy plus VCN-treated cells in acute myelocytic leukemia. Progr. Cancer Res. Ther., vol. 6, pp. 347–353 (Raven Press, New York 1979).
157 Howard, D. R.; Taylor, C. R.: An antitumor antibody in normal human serum. Oncology 37: 142–148 (1980).
158 Hughes, R. C.; Sanford, B. H.; Jeanloz, R. W.: Regeneration of the surface glycoproteins of a transplantable mouse tumor cell after treatment with neuraminidase. Proc. natn. Acad. Sci. 69: 942–945 (1972).
159 Hughes, R. C.; Palmer, P. D.; Sanford, B. H.: Factors involved in the cytotoxicity of

normal guinea-pig serum for cells of murine tumor TA3 sublines treated with neuraminidase. J. Immunol. *111:* 1071–1080 (1973).

160 Ignarro, L. J.: Regulation of lysosomal enzyme secretion: Role in inflammation. Agents and Actions *4:* 241–258 (1974).

161 Im, H. M.; Simmons, R. L.: Modification of graft-versus-host disease by neuraminidase treatment of donor cells. Transplantation *12:* 472–478 (1971).

162 Ishidate, M.; Hashimoto, Y.; Odashima, S.; Sudo, H.: Studies on acquired transplantation resistance. I. Pretreatment of Donryu rat with attenuated Yoshida sarcoma cells. Gann *56:* 13 (1965).

163 Itoh, K.; Kumagai, K.: Effect of tunicamycin and neuraminidase on the expression of Fc-IgM and -IgG receptors on human lymphocytes. J. Immunol. *124:* 1830–1836 (1980).

164 Iyer, G. Y. N.; Islam, M. F.; Qustel, J. H.: Biochemical aspects of phagocytosis. Nature *192:* 535–541 (1961).

165 Jacobsen, F.: Increase of the in vitro complement-dependent cytotoxicity against autologous invasive human bladder tumor cells by neuraminidase treatment. Acta path. microbiol. immunol. scand. Sect. C *90:* 187–192 (1982).

166 Jamieson, C. W.: Enhancement of antigenicity of syngeneic murine melanoma by neuraminidase. Int. Cancer Congress, Florence (Workshop I, Abstract): 233–234 (1974).

167 Jerne, N. K.; Nordin, A. A.; Henry, C. C.: The agar plaque technique for recognizing antibody producing cells; in Amos, Koptowski, Cell bound antibodies, pp. 109–125 (Wistar Institute Press, Philadelphia 1963).

168 Johannsen, R.; Sedlacek, H. H.: Specificity of cytotoxic antibodies to autologous human lymphocytes treated with neuraminidase from Vibrio cholerae. Behring Inst. Mitt. *55:* 209–215 (1974).

169 Johannsen, R.; Carlssohn, A. B.; Sedlacek, H. H.: In vitro transformation of human lymphocytes by neuraminidase from Vibrio cholerae (VCN) (Abstract). 6th Workshop on Leukocyte Cultures, Basel, March 17–19 (1975).

170 Johannsen, R.; Sedlacek, H. H.; Seiler, F. R.: Adjuvant effect of Vibrio cholerae neuraminidase on the in vitro and in vivo immune response; in Rainer, Immunotherapy of malignant diseases, pp. 244–259 (Schattauer, Stuttgart 1978).

171 Johannsen, R.; Sedlacek, H. H.; Schmidtberger, R.; Schick, H. J.; Seiler, F. R.: Characteristics of cytotoxic antibodies against neuraminidase-treated lymphocytes in man. J. natn. Cancer Inst. *62:* 733–742 (1979).

172 Johannsen, R.; Sedlacek, H. H.; Seiler, F. R.: The neuraminidase-induced enhanced immune response can be attributed to immune stimulatory and antigenic properties of the enzyme. Transplant. Proc. *11:* 1411–1412 (1979).

173 Jondal, M.; Holm, G.; Wigzell, H.; Surface markers on human T and B lymphocytes. I. A large population of lymphocytes forming non-immune rosettes with SRBC. J. exp. Med. *136:* 207–215 (1972).

174 Kania, J.; Uhlenbruck, G.; Janssen, E.; Klein, P. J.; Vierbuchen, M.: Isolation, characterization and implications of anti-TF (Thomson-Friedenreich) agglutinins from different sources. Immunobiol. *157:* 154–168 (1980).

175 Kassulke, J. T.; Stutman, O.; Yunis, E. J.: Blood group isoantigens in leukemic cells. Reversibility of isoantigenic change by neuraminidase. J. natn. Cancer Inst. *46:* 1201–1208 (1971).

176 Katz, A.; Klajman, A.; Yaretzky, A.; Steiner, Z. P.; Knyszynski, A.: Detection of anti red blood cell antibodies by treatment of the cells with neuraminidase. Scand. J. Haematol. *18:* 98–104 (1977).

177 Kedar, E.; Schwartzbach, M.; Raanan, U.; Hefetz, S.: In vitro induction of cell-mediated immunity to murine leukemia cells. II. Cytotoxic activity in vitro and tumor-neutralizing capacity in vivo of anti-leukemia cytotoxic lymphocytes generated in macrocultures. J. Immunol. Methods *16:* 39–58 (1977).

178 Keenan, T. W.; Morré, D. J.: Mammary carcinoma: Enzymatic block in disialoganglioside biosynthesis. Science *182:* 935–937 (1973).

179 Kemp, R. B.: Effect of the removal of cell surface sialic acids on cell aggregation in vitro. Nature, Lond. *218:* 1255–1256 (1968).

180 Kerkhaert, J. A. M.; Berg, J.; Willers, J. M. N.: Influence of cyclophosphamide on delayed type hypersensitivity in the mouse. Ann. Immun. Inst. Pasteur *125 C:* 415–426 (1974).

181 Killion, G. J.: The immunotherapeutic value of a L1210 tumour-cell vaccine depends upon the expression of cell-surface carbohydrates. Cancer Immunol. Immunother. *3:* 87–91 (1977).

182 Killion, J. J.: Immunotherapy with tumor cell subpopulations. III. Interaction between specific and nonspecific immunostimulants. Cancer Immunol. Immunother. *5:* 27–30 (1978).

183 Kim, Z.; Uhlenbruck, G.: Untersuchungen über T-Antigen und T-Agglutinin. Z. Immunitätsforsch. *130:* 88–99 (1966).

184 Klebanoff, S. J.: Oxygen intermediates and the microbicidal event; in Van Furth, Mononuclear phagocytes: Functional aspects, Part II, pp. 1106–1141 (Martinus Nijhoff, The Hague 1980).

185 Klein, M.; Neauport-Santes, C.; Ellerson, J. R.; Fridman, W. H.: Binding site of human IgG subclasses and their domains for Fc-receptors of activated murine T-cells. J. Immunol. *119:* 1077–1083 (1977).

186 Klein, P. J.; Newman, R. A.; Müller, P.; Uhlenbruck, G.; Schaefer, H. E.; Lennartz, K. J.; Fischer, R.: Histochemical methods for the demonstration of Thomsen-Friedenreich antigen in cell suspensions and tissue sections. Klin. Wschr. *56:* 761–765 (1978).

187 Klein, P. J.; Newman, R. A.; Müller, P.; Uhlenbruck, G.; Citoler, P.; Schaefer, H. E.; Lennartz, K. J.; Fischer, R.: The presence and significance of the Thomsen-Friedenreich antigen in breast tissue. II. Its topochemistry in normal, hyperplastic and carcinoma tissue of the breast. J. Cancer Res. clin. Oncol. *93:* 205–214 (1979).

188 Knop, J.: Effect of Vibrio cholerae neuraminidase on the mitogen response of T-lymphocytes. I. Enhancement of macrophage T-lymphocyte cooperation in concanavalin A-induced lymphocyte activation. Immunobiology *157:* 474–485 (1980).

189 Knop, J.: Effect of Vibrio cholerae neuraminidase on the mitogen response of T-lymphocytes. II. Modulation of the lymphocyte response to macrophage released factors by neuraminidase. Immunobiology *157:* 486–498 (1980).

190 Knop, J.; Ax, W.; Sedlacek, H. H.; Seiler, F. R.: Effect of Vibrio cholerae neuraminidase on the phagocytosis of E. coli by macrophages in vivo and in vitro. Immunology *34:* 555–563 (1973).

191 Knop, J. et al.: Stimulatory effect of Vibrio cholerae neuraminidase on the antibody response to various antigens. Immunology *34:* 181–187 (1978).

192 Kölsch, E.; Mengersen, R.: Low numbers of tumor cells suppress the host immune system. Adv. exp. med. Biol. 66: 431–436 (1976).
193 Kolb, H.; Kriese, A.; Kolb-Bachofen, V.; Kolb, H.-A.: Possible mechanism of entrapment of neuraminidase-treated lymphocytes in the liver. Cell Immunol. 40: 457–462 (1978).
194 Kolb, H.; Schudt, C.; Kolb-Bachofen, V.; Kolb, H.-A.: Cellular recognition by rat liver cells of neuraminidase-treated erythrocytes. Exp. Cell Res. 113: 319–325 (1978).
195 Kolb, H.; Vogt, D.; Herbertz, L.; Corfield, A.; Schauer, R.; Schlepper-Schäfer, J.: The galactose-specific lectins on rat hepatocytes and Kupffer cells have identical binding characteristics. Hopper-Seyler's Z. Physiol. Chem. 361: 1747–1750 (1980).
196 Kollmorgen, G. M.; Erwin, D. N.; Kollion, J. J.; Hoge, A. F.; Sansing, W. A.: Combination chemotherapy and immunotherapy of transplantable murine leukemia. Proc. Am. Ass. Cancer Res. 14: 69 (1973).
197 Kollmorgen, G. M.; Killion, J. J.; Sansing, W. A.; Cantrell, J. I.; Bundren, J. C.; LeFever, A. V.: Immunotherapy with neuraminidase-treated cells and bacillus Calmette-Guérin. Surgery 79: 202–208 (1976).
198 Kozugi, K.: Beitrag zur Statistik der Geschwülste bei den Haussäugetieren. Diss. Universität Gießen (1973).
199 Kraemer, P. M.: Regeneration of sialic acid on the surface of Chinese hamster cells in culture. I. General characteristics of the replacement process. J. Cell. Physiol. 68: 85–90 (1966).
200 Kreider, J. W.; Bartlett, G. L.; Purnell, D. M.: Inconsistent response of B16 melanoma to BCG Immunotherapy. J. natn. Cancer Inst. 56: 803–810 (1976).
201 Küster, J. M.; Schauer, R.: Phagocytosis of sialidase-treated rat erythrocytes: evidence for a two-step mechanism. Hoppe-Seyler's Z. Physiol. Chem. 362: 1507–1514 (1981).
202 Kuhn, R.; Brossmer, R.: Über O-Acetyl-Lactaminsäurelactose aus Kuhcolostrum und ihre Spaltbarkeit durch Influenzavirus. Ber. Dt. Chem. Ges. 89: 2013 (1956).
203 Lagrange, P. H.; Mackaness, G. B.; Miller, T. E.: Potentiation of T-cell mediated immunity by selective suppression of antibody formation with cyclophosphamide. J. exp. Med. 139: 1529–1539 (1974).
204 Lamelin, J.-P.; Lisowska-Bernstein, B.; Matter, A.; Ryser, J. E.; Vassalli, P.: Mouse thymus-independent and thymus-derived lymphoid cells. I. Immunofluorescent and functional studies. J. exp. Med. 136: 984–1007 (1972).
205 Landsteiner, K.; Levine, P.: On individual differences in human blood. J. exp. Med. 47: 757–775 (1928).
206 Lay, W. H.; Mendes, N. F.; Bianco, C.; Nussenzweig, V.: Binding of sheep red blood cells to a large population of human lymphocytes. Nature, Lond. 230: 531–532 (1971).
207 Lee, A.: Effect of neuraminidase on the phagocytosis of heterologous red cells by mouse peritoneal macrophages. Proc. Soc. exp. Biol. Med. 128: 891–894 (1968).
208 LeFever, A. V.; Killion, J. J.; Kollmorgen, G. M.: Active immunotherapy of L1210 leukemia with neuraminidase-treated, drug-resistant L1210 sublines. Cancer Immunol. Immunother. 1: 211–217 (1976).
209 Lin, J. S.; Murphy, W. H.: Dependence of immunity to IB leukemia on an adjuvant effect of immunizing cell preparations. Cancer Res. 29: 2163–2168 (1969).

210 Lindahl-Kiessling, K.; Peterson. R. D. A.: The mechanism of phytohemagglutinin action. III. Stimulation of lymphocytes by allogeneic lymphocytes and phytohemagglutinin. Exp. Cell Res. 55: 85–87 (1969).

211 Lindenmann, J.; Klein, P. A.: Immunological aspects of viral oncolysis. Recent Res. Cancer Res. 9: 1–84 (1967).

212 Littman, M. L.; Kim, Y. C.; Suk, D.: Immunization of mice to sarcoma 180 and Ehrlich carcinoma with ultraviolet-killed tumor vaccine. Proc. Soc. exp. Biol. Med. 127: 7–17 (1968).

213 Lüben, G.; Sedlacek, H. H.,; Seiler, F. R.: Quantitative experiments on the cell membrane binding of neuraminidase. Behring Inst. Mitt. 59: 30–37 (1976).

214 Lundgren, G.; Jeitz, L.; Lundin, L.; Simmons, R. L.: Increased stimulation by neuraminidase-treated cells in mixed lymphocyte cultures. Fed. Proc. 30: 395 (1971).

215 Lundgren, G.; Simmons, R. L.: Effect of neuraminidase on the stimulatory capacity of cells in human mixed lymphocyte cultures. Clin. exp. Immunol. 9: 915–926 (1971).

216 Luner, S. J.; Sturgeon, P.; Szklarek, D.; McQuiston, D. T.: Effects of proteases and neuraminidase on RBC surface charge and agglutination: A kinetic study. Vox. Sang. 28: 184–199 (1975).

217 Lutz, D.: Klinik und Immunologie bei akuten myeloischen Leukämien unter einer Chemoimmuntherapie. Onkologie 4: 202–212 (1981).

218 MacEwen, E. G.; Patnaik, A. K.; Harvey, H. J.; Panko, W. B.: Estrogen receptors in canine mammary tumors. Cancer Res. 42: 2255–2259 (1982).

219 Mackaness, G. B.: The immunological basis of acquired cellular resistance. J. exp. Med. 120: 105 (1964).

220 Mackaness, G. B.; Lagrange, P. H.; Miller, T. E.; Ishibashi, T.: The formation of activated T-cells; in Wagner et al., Activation of macrophages, p. 193 (Ecerpta Medica, Amsterdam 1974).

221 Maguire, H. C.; Ettore, V. I.: Enhancement of dinitrochlorobenzené (DNCB) contact sensitization by cyclophosphamide in guinea pigs. J. invest. Derm. 48: 39–43 (1967).

222 Malavé, I.; Blanca, I.; Fuji, H.: Influence of inoculation site on development of the Lewis lung carcinoma and suppressor cell activity in syngeneic mice. J. natn. Cancer Inst. 62: 83–88 (1979).

223 Mathé, G.; Halle-Pannenko, O.; Bourut, C.: Active immunotherapy in spontaneous leukemia of AKR mice. Exp. Hematol., Copenh. 1: 110–114 (1973).

224 Mayhew, E.: Cellular electrophoretic mobility and the mitotic cycle. J. Gen. Physiol. 49: 717–725 (1966).

225 Mayhew, E.; Weiss, L.: Ribonucleic acid at the periphery of different cell types and effect of growth rate on ionogenic groups in the periphery cultured cells. Exp. Cell Res. 50: 441–453 (1968).

226 McKenzie, I. F. C. et al.: Ia antigenic specificities are oligosaccharide in nature: Hapten inhibition studies. J. exp. Med. 145: 1039–1053 (1977).

227 McKolanis, J. R.; Veltri, R. W.: Antibody response to a solubilized tumor-associated membrane antigen (TAMA) from the murine Lewis lung tumor. Cancer Immunol. Immunother. 15: 131–137 (1983).

228 McQuiddy, P.; Lilien, J. E.: The binding of exogenously added neuraminidase to cells and tissues in culture. Biochim. Biophys. Acta. 291: 774–779 (1973).

229 McTaggart, S. P.; Burus, W. H.; White, D. D.; Jackson, D. C.: Fc-receptors induced by herpes simplex virus. J. Immunol. *121:* 726–730 (1978).
230 Mehrishi, J. N.; Zeiller, K.; Sachtleben, P.: Murine T and B lymphocytes: The effect of neuraminidase and anti-neuraminidase serum on the surface charge of enzyme-treated cells. Behring Inst. Mitt. *55:* 243–245 (1974).
231 Meindl, P.; Tuppy, H.: Über die Spaltung synthetischer Sialinsäure-Ketoside durch Neuraminidase. Mh. Chem. *97:* 990–999 (1966).
232 Meindl, P.; Bodo, G.; Palese, P.; Schulman, J.; Tuppy, H.: Inhibition of neuraminidase activity by derivatives of 2-deoxy-2,3-dehydro-N-acetylneuraminic acid. Virology *58:* 457–463 (1974).
233 Metzger, R. S.; Mohanakumar, T.; Miller, D. S.: Antigens specific for human lymphocytic and myeloid leukemia cells: Detection of nonhuman primate antisera. Science *178:* 986–988 (1972).
234 Miller, C. W.; DeBlasi, R. F.; Fisher, S. J.: Immunological studies in murine osteosarcoma. J. Bone Joint. Surg. (Am.) *58:* 312–319 (1976).
235 Mills, G.; Monticone, V.; Paetkau, V.: The role of macrophages in thymocyte mitogenesis. J. Immunol. *117:* 1325–1330 (1976).
236 Mobley, J. R.; Graber, C. D.; O'Brien, P. H.; Gale, G. R.; Neeley, B.: Phytomitogen and neuraminidase in the treatment of Ehrlich carcinoma in mice. Res. Commun. chem. pathol. Pharmacol. *9:* 155–162 (1974).
237 Mohr, E.; Schramm, G.: Reinigung und Charakterisierung der Neuraminidase aus Vibrio cholerae. Z. Naturforsch. *15B:* 568–575 (1960).
238 Müller, H. E.: Neuraminidase of bacteria and protozoa and their pathogenic role. Behring Inst. Mitt. *55:* 34–56 (1974).
239 Murphy, W. H.; Herlocher, M. L.; Griep, J.: Immunization of C58 mice to line I_b leukemia. J. infect. Dis. *112:* 28–36 (1963).
240 Nakano, T.; Imai, Y.; Sawada, J.; Osawa, T.: The use of various lectins for the separation of T-cell subsets; in Schauer, Glycoconjugates. Proc. 5th Int. Symp. Kiel, Sept. 1979, p. 452 (Thieme, Stuttgart 1979).
241 Naor, D.: Suppressor cells: Permitters and promoters of malignancy? Adv. Cancer Res. *29:* 45–125 (1979).
242 Nelson, L. W.; Carlton, W. W.; Weikle, J. H., Jr.: Canine mammary neoplasms and progestogens. JAMA *219:* 1601–1606 (1972).
243 Neurath, A. R.; Hartzell, R. W.; Rubin, B. R.: Isoelectric focussing of neuraminidase. Experientia *26:* 1210–1211 (1970).
244 Newman, R. A.; Glöckner, W. M.; Uhlenbruck, G. G.: Immunochemical detection of the Thomsen-Friedenreich antigen (T-antigen) on the pig lymphocyte plasma membrane. Eur. J. Biochem. *64:* 373–380 (1976).
245 Newman, R. A.; Harrison, R.; Uhlenbruck, G.: Alkali-labile oligosaccharide form bovine milk fat globule membrane glycoprotein. Biochim. Biophys. Acta *433:* 344–356 (1976).
246 Newman, R. A.; Uhlenbruck, G.: Investigation into the occurrence and structure of lectin receptors on human and bovine erythrocyte, milk fat globule and lymphocyte plasma membrane glycoproteins. Eur. J. Biochem. *76:* 149–155 (1977).
247 Newman, R. A. et al.: Binding of peanut lectin to breast epithelium, human carcinomas, and a cultured rat mammary stem cell: Use of the lectin as a marker of mammary differentiation. J. natn. Cancer Inst. *63:* 1339–1346 (1979).

248 Newman, R. A.; Klein, P. J.; Uhlenbruck, G.; Citoler, P.; Karduck, D.: The presence and significance of the Thomsen-Friedenreich antigen in breast cancer. I. Serological studies. J. Cancer Res. clin. Oncol. *93:* 181–188 (1979).

249 Nicolson, G. L.; Poste, G.: The cancer cell: dynamic aspects and modification in cell-surface organization (first of two parts). New Engl. J. Med. *295:* 197–203 (1976).

250 Noonan, K. D.: Lectins as probes of the transformed cell surface; in Nicolau, Virus-transformed cell membranes, pp. 281–371 (Academic Press, New York 1978).

251 Nordling. S. E.; Mayhew, E.: On the intracellular uptake of neuraminidase. Exp. Cell Res. *44:* 552–562 (1966).

252 Nordt, F. J.; Franco, M.; Corfield, A.; Schauer, R.; Ruhenstroth-Bauer, G.: The efficacy of neuraminidase treatment in studies on red cell aging. Blut *42:* 95–98 (1981).

253 Noseworthy, J., Jr.; Korchak, H.; Karnovsky, M. L.: Phagocytosis and the sialic acid of the surface of polymorphonuclear leukocytes. J. Cell Physiol. *79:* 91–96 (1972).

254 Novogrodsky, A.: Potentiation of phytomitogen action by neuraminidase and basic polypeptides. Nature *250:* 788–790 (1974).

255 Nowak, J. S.: Effect of neuraminidase treatment on the expression of Fc-IgG receptors on chicken embryonic bursa and thymus cells. Immunol. Lett. *10:* 141–144 (1985).

256 Ohta, N.; Pardee, A. B.; McAuslan, B. R.; Burger, M. M.: Sialic acid contents and controls of normal and malignant cells. Biochim. Biophys. Acta *158:* 98–102 (1968).

257 Okuda, K.; Cullen, S. E.; Hilgers, J.; David, C.: Immune response-associated antigens on mouse leukemia cells. I. Detection of Ia antigens on GRSL cells. Transplantation *26:* 153–156 (1978).

258 Ortiz-Ortiz, L.; Nakamura, R. M.; Weigle, W. O.: T cell requirement for experimental allergic encephalomyelitis induction in the rat. J. Immunol. *117:* 576–577 (1976).

259 Pape, G. R.; Troye, M.; Perlmann, P.: Characterization of cytolytic effector cells in peripheral blood of healthy individuals and cancer patients. I. Surface markers and K cell activity after separation of B cells and lymphocytes with Fc-receptors by column fractionation. J. Immunol. *118:* 1919–1924 (1977).

260 Parish, C. R.; Jackson, D. C.; McKenzie, I. F. C.: Low-molecular weight Ia antigens in normal mouse serum. III. Isolation and partial chemical characterization. Immunogenetics *3:* 455–463 (1976).

261 Parish, C. R.; Higgins, T. J.; McKenzie, I. F. C.: Comparison of antigens recognized by xenogeneic and allogeneic anti-Ia antibodies: Evidence for two classes of Ia antigens. Immunogenetics *6:* 343–354 (1978).

262 Parish, C. R.; McKenzie, I. F. C.: A detailed serological analysis of a xenogeneic anti-Ia serum. Immunogenetics *6:* 183–196 (1978).

263 Parish, C. R.; Higgins, T. J.; McKenzie, I. F. C.: Lymphocytes express Ia antigens of foreign haplotype following treatment with neuraminidase. Immunogenetics *12:* 1–20 (1981).

264 Paterson, P. Y.; Hanson, M. A.: Cyclophosphamide inhibition of experimental allergic encephalomyelitis and cellular transfer of the disease in Lewis rats. J. Immunol. *103:* 1311–1316 (1969).

265 Paterson, P. Y.; Drobih, D. G.; Hanson, M. A.; Jacobs, A. F.: Induction of experimental allergic encephalomyelitis in Lewis rats. Int. Archs Allergy appl. Immun. *37:* 26–40 (1970).

266 Pauly, J. L.; Germain, M. J.; Han, T.: Neuraminidase alteration of human lymphocyte reactivity to mitogens, antigens and allogeneic lymphocytes. J. Med. 9: 223–236 (1978).

267 Pereira, M. E. A.; Kabat, E. A.; Lotan, R.; Sharon, N.: Immunochemical studies on the specificity of the peanut (Arachis hypogaea) agglutinin. Carbohydr. Res. 51: 107–118 (1976).

268 Petitou, M.; Rosenfeld, C.; Sinay, P.: A new assay for cell-bound neuraminidase. Cancer Immunol. Immmunother. 2: 135–137 (1977).

269 Pfreundschuh, M.; Dörken, B.; Brandeis, W.; Hunstein, W.; Wernet, P.: Effect of neuraminidase treatment on serum reactivity to autologous leukemic blast cells. Cancer Immunol. Immunother. 15: 194–199 (1983).

270 Pielken, H.-J.; Urbanitz, D.; Büchner, T.; Hiddemann, W.; Koch, P.; Van de Loo, J.: Immunotherapy with neuraminidase-treated allogeneic blasts in the treatment of acute myeloid leukemia. Preliminary results of a randomized study. Dt. Krebskongress, München, März 1984.

271 Pimm, M. V.; Cook, A. J.; Baldwin, R. W.: Failure of neuraminidase treatment to influence tumorigenicity or immunogenicity of syngeneically transplanted rat tumor cells. Eur. J. Cancer 14: 869–878 (1978).

272 Pincus, J. H.; Jameson, A. K.; Brandt, A. E.: Immunotherapy of L1210 leukemia using neuraminidase-modified plasma membranes combined with chemotherapy. Cancer Res. 41: 3082–3086 (1981).

273 Ploemacher, R. E.; Brons, N. H. C.; De Vreede, E.; Van Soest, P. L.: Colony formation by bone marrow cells after incubation with neuraminidase. II. Sensitivity of erythroid progenitor cells for burst promoting activity and erythropoietin and restoration of reduced spleen colony formation in mice pretreated with desialated erythrocyte membrane fragments. Exp. Hematol. 9: 156–167 (1981).

274 Porwit-Bobr, Z.; Slowik, M.; Tomecki, J.: Effect of neuraminidase-treated and mitomycin C-treated polyoma tumor cells on the established tumor growth in CBA mice. I. An attempt at evaluation of polyoma tumor destruction using the distribution of lissamine green method. Folia Histochem. Cytochem. 12: 311320 (1974).

275 Prager, M. D.; Ribble, R. J.; Mehta, J. M.: Aspects of the immunology of the tumor-host relationship and responsiveness to modified lymphoma cells, Int. Symp. on Investigation and Stimulation of Immunity in Cancer Patients, CNRA Paris, pp. 379–386 (1972).

276 Prager, M. D.; Ticaric, S.; Merrill, C. L.: Tumor host relationship on immune response to modified lymphoma cells. Proc. Am. Ass. Cancer Res. 13: 103 (1972).

277 Prager, M. D.; Hollinshead, A. C.; Ribble, R. J.; Derr, I.: Induction of immunity to a mouse lymphoma by multiple methods, including vaccination with soluble membrane fractions. J. natn. Cancer Inst. 51: 1603–1607 (1973).

278 Presant, C. A.; Kornfeld, S.: Characterization of the cell surface receptor for the Agaricus bisporus haemagglutinin. J. biol. Chem. 247: 6937–6945 (1972).

279 Prokop, O.; Uhlenbruck, G.; Kohler, W.: A new source of antibody-like substances having anti blood group specificity. Vox. Sang. 14: 321–333 (1968).

280 Rahman, A. F. R.; Longenecker, B. M.: A monoclonal antibody specific for the Thomsen-Friedenreich cryptic T antigen. J. Immunol. 129: 2021–2024 (1982).

281 Rainer, H.; Kovats, E.; Lehmann, H. G.; Micksche, M.; Rauhs, R.; Sedlacek, H. H.; Seidl, W.; Schemper, M.; Schiessel, R.; Schweiger, B.; Wunderlich, M.: Effectiveness

of postoperative adjuvant therapy with cytotoxic chemotherapy (cytosine arabinoside, mitomycin C, 5-fluorouracil) or immunotherapy (neuraminidase-modified allongeneic cells) in the prevention or recurrence of Duke's B and C colon cancer. Recent Res. Cancer Res. 79: 41–47 (1981).

282 Rainer, H.; Dittrich, C.; Rauhs, R.; Schemper, M.; Schiessel, R.; Wunderlich, M.; Micksche, M.; Vetterlein, M.; Kovats, E.; Lehmann, H. G.; Sedlacek, H. H.: Effectiveness of postoperative adjuvant therapy with cytotoxic chemotherapy (cytosine arabinoside, motomycin C, 5-fluorouracil) or immunotherapy (neuraminidase-modified allogeneic cells) in the prevention or recurrence of Duke's B and C colon cancer; in Jones, Salmon, Adjuvant therapy of cancer, vol. II, pp. 479–485 (Grune & Stratton, New York 1984).

283 Rao, V. S.; Bennett, J. A.; Shen, F. W.; Gershon, R. K.; Mitchell, M. S.: Antigen-antibody complexes generate Lyt 1 inducers of suppressor cells. J. Immunol. 125: 63–67 (1980).

284 Ray, P. K.; Gewurz, H.; Simmons, R. L.: The mechanism of increased sensitivity of neuraminidase-treated cells to antibody-induced cytolysis. Fed. Proc. 29: 573 (1970).

285 Ray, P. K.; Gewurz, H.; Simmons, R. L.: Complement sensitivity of neuraminidase-treated lymphoid cells. Transplantation 12: 327–329 (1971).

286 Ray, P. K.; Simmons, R. L.: Failure of neuraminidase to unmask alloageneic antigens on cell surface. Proc. Soc. exp. Biol. Med. 138: 600–604 (1971).

287 Ray, P. K.; Gewurz, H.; Simmons, R. L.: The serologic behaviour of neuraminidase-treated lymphoid cells: Alloantigenicity and complement sensitivity. Clin. exp. Immunol. 11: 441–460 (1972).

288 Ray, P. K.; Simmons, R. L.: Comparative effect of viral and bacterial neuraminidase on the complement sensitivity of lymphoid cells. Clin. exp. Immunol. 10: 139–150 (1972).

289 Ray, P. K.; Simmons, R. L.: Differential release of sialic acid from normal and malignant cells by Vibrio cholerae neuraminidase or influenza virus neuraminidase. Cancer Res. 33: 936–939 (1973).

290 Ray, P. K.; Sundaram, K.: Neuraminidase-induced immunotherapy of cancer, Int. Cancer Congress, Florence, p. 472 (1974).

291 Ray, P. K.; Chatterjee, S.: Neuraminidase treatment enhances the lysolecithin-induced intercellular adhesion of Amoeba proteus. Z. Naturf. (C) 30: 551–552 (1975).

292 Ray, P. K.; Thakur, V. S.; Sundaram, K.: Antitumor immunity. I. Differential response of neuraminidase-treated and x-irradiated tumor vaccine. Eur. J. Cancer 11: 1–8 (1975).

293 Ray, P. K.; Thakur, V. S.; Sundaram, K.: Antitumor immunity. II. Viability, tumorgenicity and immunogenicity of neuraminidase-treated tumor cells: Effective immunization of animals with a tumor vaccine. J. natn. Cancer Inst. 56: 83–87 (1976).

294 Ray, P. K.; Seshadri, M.: Inhibition of growth of rat Yoshida sarcoma using a neuraminidase-treated tumour vaccine. Indian J. exp. Biol. 17: 36–39 (1979).

295 Raz, A.; McLellan, W. L.; Hart, I. R.; Bucana, C. D.; Hoyer, L. C.; Sela, B.-A.; Dragsten, P.; Fidler, I. J.: Cell surface: properties of B16 melanoma variants with differing metastatic potential. Cancer Res. 40: 1645–1651 (1980).

296 Reed, R. C.; Gutterman, J. U.; Mavligit, G. M.; Hersh, E. M.: Sialic acid on leuke-

mia cells: Relation to morphology and tumor immunity. Proc. Soc. exp. Biol. Med. *145:* 790–793 (1974).
297 Reisner, E. G.: Relation of the neuraminidase-revealed antigens of human lymphocytes and erythrocytes. Transplantation *19:* 357–360 (1975).
298 Reisner, E. G.; Amos, D. B.: The complement-binding and absorptive capacity of human white blood cells treated with neuraminidase. Transplantation *14:* 455–461 (1972).
299 Reisner, E. G.; Flye, K.; Su Chung, S.; Amos, D. B.: The cytotoxic reactivity and sialic acid content of human lymphoid cells. Tissue Antigens *4:* 7–20 (1974).
300 Reisner, Y.; Linker-Israeli, M.; Sharon, N.: Separation of mouse thymocytes into two subpopulations by the use of peanut agglutinin. Cell Immunol. *25:* 129–134 (1976).
301 Rhodes, J.: Receptor for monomeric IgM on guinea pig splenic marcophages. Nature, Lond. *243:* 527–528 (1973).
302 Rios, A.; Simmons, R. L.: Comparative effect of mycobacterium Bovis- and neuraminidase-treated tumor cells on the growth of established methylcholanthrene fibrosarcomas in syngeneic mice. Cancer Res. *32:* 16–21 (1972).
303 Rios, A.; Simmons, R. L.: Immunospecific regression of various syngeneic mouse tumors in response to neuraminidase-treated tumor cells. J. natn. Cancer Inst. *51:* 637–644 (1973).
304 Rios, A.; Simmons, R. L.: Active specific immunotherapy of minimal residual tumor: Excision plus neuraminidase-treated tumor cells. Int. J. Cancer *13:* 71–81 (1974).
305 Rios, A.; Simmons, R. L.: Comparative tumor immunoregressive effect of neuraminidase concanavalin A or irradiated tumor cells. Fed. Proc. *33:* 615 (1974).
306 Röllinghoff, M.; Starzinski-Powitz, A.; Pfizenmaier, K.; Wagner, H.: Cyclophosphamide-sensitive T lymphocytes suppress the in vivo generation of antigen specific cytotoxic T lymphocytes. J. exp. Med. *145:* 455–459 (1977).
307 Rogentine, G. N., r.: Naturally occurring human antibody to neuraminidase-treated human lymphocytes. Antibody levels in normal subjects, cancer patients, and subjects with immunodeficiency. J. natn. Cancer Inst. *54:* 1307–1311 (1975).
308 Rogentine, G. N., Jr.; Plocinik, B. A.: Carbohydrate inhibition studies of the naturally occurring human antibody to neuraminidase-treated human lymphocytes. J. Immunol. *113:* 848–858 (1974).
309 Rogentine, G. N., Jr.; Doherty, C. M.; Pincus, S. H.: Increase in titer of the naturally occurring human antibody to neuraminidase-treated lymphocytes after influenza. J. Immunol. *119:* 1652–1654 (1977).
310 Ronneberger, H.: Toxicological studies with therapeutically applicable Vibrio cholerae neuraminidase. Devl. Biol. Stand. *38:* 413–419 (1978).
311 Rosato, F. E.; Brown, A. S.; Miller, E. E.; Rosato, E. F.; Mullis, W. F.; Johnson, J.; Moskovitz, A.: Neuraminidase immunotherapy of tumours in man. Surg. Gynaec. Obstet. *139:* 675–682 (1974).
312 Rosato, R. E.; Miller, E.; Rosato, E.; Brown, A.; Wallack, M.; Johnson, J.; Moskovitz, A.: Active specific immunotherapy of human solid tumours. Ann. N. Y. Acad. Sci. *277:* 332–338 (1976).
313 Rosenberg, S. A.; Einstein, A. B.: Sialic acids on the plasma membrane of cultured human lymphoid cells: Chemical aspects and biosynthesis. J. Cell. Biol. *53:* 466–473 (1972).

314 Rosenberg, S. A.; Plocinik, B. A.; Rogentine, G. N.: 'Unmasking' of human lymphoid cell heteroantigens by neuraminidase treatment. J. natn. Cancer Inst. *48:* 1271–1276 (1972).
315 Rosenberg, S. A.; Rogentine, G. N., Jr.: Natural human antibodies to 'hidden' membrane components. Nature new Biol. *239:* 203–204 (1972).
316 Rosenberg, S. A.; Schwarz, S.: Murine antibodies to a cryptic membrane antigen: Possible explanation for neuraminidase-induced increase in cell immunogenicity. J. natn. Cancer Inst. *52:* 1151–1155 (1974).
317 Rosenberg, S. A.; Schwartz, S.; Baker, A.: Natural antibodies to 'cryptic' membrane antigens exposed by treatment with neuraminidase. Behring Inst. Mitt. *55:* 204–208 (1974).
318 Rossi, F.; Romeo, D.; Patriarca, P.: Mechanism of phagocytosis-associated oxidative metabolism in polymorphonuclear leukocytes and macrophages. J. Reticuloendothel. Soc. *12:* 127–149 (1972).
319 Rothauge, C. F.; Kraushaar, J.: Immunotherapie metastasierender Prostata- und hypernephroider Nierenkarzinome. Diagnostik *17:* 22–26 (1984).
320 Rowley, D.: Sensitivity of rough gram-negative bacteria to the bactericidal action of serum. J. Bact. *95:* 1647–1650 (1968).
321 Rowley, P. T.; Ohlsson-Wilhelm, B. M.; Farley, B. A.: Erythropoiesis in vitro: enhancement by neuraminidase. Blood *57:* 483–492 (1981).
322 Rühl, H.; Fülle, H. H.; Kopeen, K. M.; Schwerdtfeger, R.: Adjuvant specific immunotherapy in maintenance treatment of adult acute non-lymphocytic leukaemia. Klin. Wschr. *59:* 1189–1193 (1981).
323 Ruhenstroth-Bauer, G. et al.: Elektrophoretische Untersuchungen an normalen und malignen Zellen. Naturwissenschaften *49:* 363–368 (1962).
324 Ruhenstroth-Bauer, G.; Fuhrmann, G. F.; Kübler, W.; Rueff, F.; Munk, K.: Zur Bedeutung der Neuraminsäuren in der Zellmembran für das Wachstum maligner Zellen. Z. Krebsforsch. *65:* 37 (1962).
325 Ryan, G. B.; Majno, G.: Acute inflammation. Am. J. Path. *86:* 183–276 (1977).
326 Sachtleben, P.; Gsell, R.; Mehrishi, J. N.: Neuraminidase and anti-neuraminidase serum: effect on the cell surface properties. Vox Sang. *25:* 519–528 (1973).
327 Sakano, T.; Kittaka, E.; Hyodo, S.; Horino, N.; Tanaka, Y.; Usui, T.: Effect of neuraminidase on the duration of contact with con A in lymphocyte stimulation. Hiroshima J. Med. Sci. *30:* 161–163 (1981).
328 Saluja, P. G.; Hamilton, J. M.; Gronow, M.; Misdorp, W.: Pituitary prolactin levels in canine mammary cancer. Eur. J. Cancer *10:* 63–66 (1974).
329 Sanford, B. H.: An alteration in tumour histocompatibility induced by neuraminidase. Transplantation *5:* 1273–1279 (1967).
330 Sanford, B. H.; Codington, J. F.: Alteration of the tumour cell surface by neuraminidase. Transplant. Proc. *3:* 1155–1157 (1971).
331 Sanford, B. H.; Codington, J. F.: Further studies on the effect of neuraminidase on tumor cell transplantability. Tissue Antigens *1:* 135–161 (1971).
332 Sauter, C.; Lindenmann, J.; Gerber, A.: Agglutinination of leukemic myeloblasts by neuraminidase. Eur. J. Cancer *8:* 451–453 (1972).
333 Sbarra, A. J.; Karnovsky, M. L.: The biochemical basis of phagocytosis. I. Metabolic changes during the ingestion of particles by polymorphonuclear leukocytes. J. biol. Chem. *234:* 1355–1362 (1959).

334 Schick, H.-J., Schmidtberger, R.: Affinity of Vibrio cholerae neuraminidase to different human sialoglycoproteins. Behring Inst. Mitt. 55: 123–128 (1974).

335 Schick, H. J.; Zilg, H.: Production and quality control of therapeutically applicable Vibrio cholerae neuraminidase (VCN). Devl. Biol. Standard. 38: 81–85 (1975).

336 Schlesinger, M.; Amos, D. B.: The effect of neuraminidase on the serological properties of murine lymphoid cells. Transplant. Proc. 3: 895–897 (1971).

337 Schlesinger, M.; Bekesi, J. G.: Natural autoantibodies cytotoxic for thymus cells and for neuraminidase-trated leukemia cells in the sera of normal AKR mice. J. natn. Cancer Inst. 59: 945–950 (1977).

338 Schmidtke, J. R.; Simmons, R. L.: Concurrent increased immunogenicity and macrophage handling of neuraminidase-treated sheep red blood cells. Fed. Proc. 32: 1015 (1973).

339 Schmidtke, J. R.; Simmons, R. L.: Augmented uptake of neuraminidase-treated sheep red blood cells: Participation of opsonic factors. J. natn. Cancer Inst. 54: 1379–1384 (1975).

340 Schmidtke, J. R.; Ferguson, R. M.; Simmons, R. L.: Effect of X-irradiation and neuraminidase on the capacity of human lymphocytes to form rosettes with sheep red blood cells. Transplantation 22: 635–638 (1976).

341 Schneider, R.; Dorn, C. R.; Taylor, D. O.: Factors influencing canine mammary cancer development and post-surgical survival. J. natn. Cancer Inst. 43: 1249–1261 (1969).

342 Schneider, D. R.; Sedlacek, H. H.; Seiler, F. R.: Studies on cell-bound Vibrio cholerae neuraminidase activity; in Schauer, Boer, Buddecke, Kramer, Vliegenthart, Wiegandt, Glycoconjugates, pp. 354–355 (Thieme, Stuttgart 1979).

343 Schorlemmer, H. U.; Bosslet, K.; Dickneite, G.; Lüben, G.; Sedlacek, H. H.: Studies on the mechanisms of action of the immunomodulator Bestatin in various screening test systems. Behring Inst. Mitt. 74: 157–173 (1984).

344 Schulof, R. S.; Fernandes, G.; Good, R. A.; Gupta, S.: Neuraminidase treatment of human T lymphocytes effect on Fc receptor phenotype and function. Clin. exp. Immunol. 40: 611–619 (1980).

345 Schwartz, A.; Askenase, P. W.; Gershon, R. K.: Regulation of delayed-type hypersensitivity reactions by cyclophosphomide-sensitive T cells. J. Immunol. 121: 1573–1577 (1978).

346 Schwartz, B. D.; Cullen, S. E.: Chemical characteristics of Ia antigens. Springer Semin. Immunopathol. 1: 85–109 (1978).

347 Scorza Smeraldi, R.; Sabbadini-Villa, M. G.; Perussia, B.; Fabio, G.; Casali, P.; Rugarli, C.: Expression of Fc and complement receptors on T-lymphocytes activated in vitro by phytohaemagglutinin (PHA). Immunology 32: 827–830 (1977).

348 Sedlacek, H. H.: Pathophysiological aspects of immune complex diseases, part I. Klin. Wschr. 58: 543–550 (1980).

349 Sedlacek, H. H.: Pathophysiological aspects of immune complex diseases, part II. Klin. Wschr. 58: 593–605 (1980).

350 Sedlacek, H. H.; Seiler, F. R.: Demonstration of Vibrio cholerae neuraminidase (VCN) on the surface of VCN-treated cells. Behring Inst. Mitt. 55: 254–257 (1974).

351 Sedlacek, H. H.; Seiler, F. R.: Dose dependency of the effect of syngeneic, Vibrio cholerae neuraminidase treated mastocytoma cells on the life span of DBA f/2 mice. Behring Inst. Mitt. 55: 343–348 (1974).

352 Sedlacek, H. H.; Meesmann, H.; Seiler, F. R.: Regression of spontaneous mammary tumors in dogs after injection of neuraminidase-treated tumor cells. Int. J. Cancer *15:* 409–416 (1975).

353 Sedlacek, H. H.; Dersjant, H.; Baudner, S.; Seiler, F. R.: Critical evaluation of the value of immunofluorescence methods for the demonstration of immunoglobulins on the lymphocyte surface. Behring Inst. Mitt. *59:* 38–63 (1976).

354 Sedlacek, H. H.; Seiler, F. R.: The effect of Vibrio cholerae neuraminidase (VCN) on the local and humoral immunological reaction of mice against SRBC. Z. ImmunForsch. *153:* 353–354 (1977).

355 Sedlacek, H. H.; Johannsen, R.; Seiler, F. R.: Possible immunological action of Vibrio cholerae neuraminidase (VCN) in tumor immunotherapy. Dev. biol. Standard. *38:* 387–398 (1978).

356 Sedlacek, H. H.; Seiler, F. R.: Effect of Vibrio cholerae neuraminidase on the cellular immune response in vivo; in Rainer, Immunotherapy of malignant diseases, pp. 268–278 (Schattauer, Stuttgart 1978)

357 Sedlacek, H. H.; Seiler, F. R.: Immunotherapy of neoplastic diseases with neuraminidase: Contradictions, new aspects, and revised concepts. Cancer Immunol. Immunother. *5:* 153–163 (1978).

358 Sedlacek, H. H.; Seiler, F. R.: Spontaneous mammary tumors in mongrel dogs. A relevant model to demonstrate tumor therapeutical success by application of neuraminidase-treated tumor cells. Dev. biol. Standard. *38:* 399–412 (1978).

359 Sedlacek, H. H.; Weise, G.; Lemmer, A.; Seiler, F. R.: Immunotherapy of spontaneous mammary tumors in mongrel dogs with autologous tumor cells and neuraminidase. Cancer Immunol. Immunother. *6:* 47–58 (1979).

360 Sedlacek, H. H.; Bengelsdorff, H. J.; Seiler, F. R.: Minimal residual disease may be treated by chessboard vaccination with Vibrio cholerae neuraminidase (VCN) and tumour cells; in Hellmann, Hilgard, Eccles, Metastasis: Clinical and experimental aspects, pp. 310–314 (Martinus Nijhoff, The Hague 1980).

361 Sedlacek, H. H.; Wanger, R.; Bengelsdorff, H.-J.: Influence of various immunomodulators on the induction of experimental autoimmune encephalomyelitis in lewis rats. Immunobiol. *157:* 99–108 (1980).

362 Sedlacek, H. H.; Hagmayer, G.; Seiler, F. R.: Tumor immunotherapy using the adjuvant neuraminidase: preliminary results of a randomized prospective study in the canine mammary tumor; in Yamamura, Kotani, Immunomodulation by microbiological products and related synthetic compounds. Excerpta Med. Int. Congr. Ser., No. 563, pp. 524–527 (1981).

363 Sedlacek, H. H.; Dickneite, G.; Schorlemmer, H. U.: Präklinische Prüfung von Immunmodulatoren unter spezieller Berücksichtigung von Chemoimmuntherapeutika. Aktuelle Onkologie, vol. 24, pp. 52–78 (Zuckschwerdt, München 1985).

364 Sedlacek, H. H.; Weidmann, E.; Seiler, F. R.: Tumour immunotherapy using Vibrio cholerae neuraminidase (VCN); in Roszkowski, Jeljaszewics, Pulverer, Bacteria and cancer, pp. 263–290 (Academic Press, London (1982).

365 Sedlacek, H. H.; Hagmayer, G.; Seiler, F. R.: Tumor therapy of neoplastic diseases with tumor cells and Neuraminidase: Further experimental studies on chessboard vaccination. Cancer Immunol. Immunother. *23:* 192–199 (1986).

366 Seigler, H. F.; Shingleton, W. W.; Metzgar, R. S.; Buckley, C. E.; Bergoc, P. M.;

Miller, D. S.; Fetter, B. F.; Phaup, M. B.: Non-specific and specific immunotherapy in patients with melanoma. Surgery 72: 162–174 (1972).
367 Seiler, F. R.; Sedlacek, H. H.; Kanzy, E. J.; Lang, W.: Über die Brauchbarkeit immunologischer Nachweismethoden zur Differenzierung funktionell verschiedener Lymphozyten: Spontanrosetten, Komplementrezeptorrosetten und Immunglobulinrezeptoren. Behring Inst. Mitt. 52: 26–72 (1972).
368 Sedlacek, H. H.: Tumorimmunologie und Tumortherapie – Eine Standortbestimmung. Beitr. Onkol., vol. 25 (Karger, Basel 1987).
369 Sciler, F. R.; Sedlacek, H. H.: Alterations of immunological phenomena by neuraminidase: Marked rise in the number of lymphocytes forming rosettes or bearing immunoglobulin receptors. Behring Inst. Mitt. 55: 258–271 (1974).
370 Seiler, F. R.; Sedlacek, H. H.; Lüben, G.; Wiegandt, H.: Alteration of the lymphocyte surface by Vibrio cholerae neuraminidase, gangliosides and lysolecithins. Behring Inst. Mitt. 59: 22–29 (1976).
371 Seiler, F. R.; Sedlacek, H. H.: Chessboard vaccination: A pertinent approach to immunotherapy of cancer with neuraminidase and tumor cells? In Rainer, Immunotherapy on malignant diseases, pp. 479–488 (Schattauer, Stuttgart 1978).
372 Seiler, F. R.; Sedlacek, H. H.: BCG versus VCN: The antigenicity and the adjuvant effect of both compounds. Recent Results Cancer Res. 75: 53–60 (1980).
373 Sethi, K. K.; Brandis, H.: Neuraminidase-induced loss in the transplantability of murine leukaemia L1210, induction of immunoprotection and the transfer of induced immunity to normal DBA/2 mice by serum and peritoneal cells. Br. J. Cancer 27: 106–113 (1973).
374 Sethi, K. K.; Brandis, H.: Synergistic cytotoxic effect of macrophages and normal mouse serum on neuraminidase-treated murine leukaemia cells. Eur. J. Cancer 9: 809–819 (1973).
375 Sethi, K. K.; Brandis, H.: Brief communication: Protection of mice from malignant tumor implants by enucleated tumor cells. J. natn. Cancer Inst. 53: 1175–1176 (1974).
376 Shaw, S.; Pichler, W. J.; Nelson, D. L.: Fc receptors on human T-lymphocytes. III. Characterization of subpopulations involved in cell-mediated lympholysis and antibody-dependent cellular cytotoxicity. J. Immunol. 122: 599–604 (1979).
377 Silverstein, S. C.; Steinman, R. M.; Cohn, Z. A.: Endocytosis. Ann. Rev. Biochem. 46: 669–772 (1977).
378 Simmons, R. L.; Rios, A.; Ray, P. K.: Mechanism of neuraminidase-induced antigen 'unmasking'. Surg. Forum 21: 265–277 (1970).
379 Simmons, R. L.; Lipschultz, M. L.; Rios, A.; Ray, P. K.: Failure of neuraminidase to unmask histocompatibility antigens on throphoblast. Nature new Biol. 231: 111–112 (1971).
380 Simmons, R. L.; Rios, A.: Immunotherapy of cancer: Immunospecific rejection of tumours in recipients of neuraminidase-treated tumour cells plus BCG. Science 174: 591–593 (1971).
381 Simmons, R. L.; Rios, A.; Lundgren, G.; Ray, P. K.; McKhann, C. F.; Haywood, G.: Immunospecific regression of methylcholanthrene fibrosarcoma using neuraminidase. Surgery 70: 38–46 (1971).
382 Simmons, R. L.; Rios, A.; Ray, P. K.: Immunogenicity and antigenicity of lymphoid cells treated with neuraminidase. Nature new Biol. 231: 179–181 (1971).

383 Simmons, R. L.; Rios, A.; Ray, P. K.; Lundgren, G.: Effect of neuraminidase on growth of a 3-methylcholanthrene-induced fibrosarcoma in normal and immunosuppresssed syngeneic mice. J. natn. Cancer Inst. 47: 1087–1094 (1971).

384 Simmons, R. L.; Rios, A.; Kersey, J. H.: Regression of spontaneous mammary carcinomas using direct injections of neuraminidase and BCG. J. Surg. Res. 12: 57–61 (1972).

385 Simmons, R. L., Rios, A.: Differential effect of neuraminidase on the immunogenicity of viral associated and private antigens of mammary carcinomas. J. Immunol. 111: 1820–1825 (1973).

386 Simmons, R. L.; Rios, A.: Cell surface modification in the treatment of experimental cancer neuraminidase or concavalin A. Cancer 34: 1541–1547 (1974).

387 Simmons, R. L.; Rios, A.: Immunospecific regression of methylcholanthrene fibrosarcoma with the use of neuraminidase. V. Quantitative aspects of the experimental immunotherapeutic model. Israel J. med. Scis 10: 925–938 (1974).

388 Simmons, R. L.; Rios, A.: Modified tumor cells in the immunotherapy of solid mammary tumors. Med. Clins N. Am. 60: 551–565 (1976).

389 Simmons, R. L.; Aranha, G. V.; Gunnarson, A.; Grage, T. B.; McKhann, C. F.: Active specific immunotherapy for advanced melanoma utilizing neuraminidase-treated autochthonous tumor cells; in Terry, Windhorst, Immunotherapy of cancer research and therapy, vol. 6, pp. 123–133 (Raven Press, New York 1978).

390 Singer, J. A.; Morrison, M.: Relative susceptibilities to neuraminidase of the glycoproteins of the human erythrocyte. Biochim. Biohpys. Acta 343: 598–608 (1974).

391 Small, M.; Trainin, N.: Separation of populations of sensitized lymphoid cells into fractions inhibiting and fractions enhancing syngeneic tumor growth in vivo. J. Immunol. 117: 292–297 (1976).

392 Smith, D. F.; Walborg, E. F.: Isolation and chemical characterization of cell surface sialoglycopeptide fractions during progression of rat ascites hepatoma AS-30D. Cancer Res. 32: 543–549 (1972).

393 Smith, B. A.; Ware, B. R.; Yankee, R. A.: Electrophoretic mobility distributions of normal human T and B lymphocytes and of peripheral blood lymphoblasts in acute lympholytic leukemia: effects of neuraminidase and of solvent ionic strength. K. Immunol. 120: 921–926 (1978).

394 Smyth, H.; Farrell, D. J.; O'Kennedy, R.; Corrigan, A.: Transplantability of neuraminidase-treated ascites tumour cells. Eur. J. Cancer 13: 1313–1320 (1977).

395 Song, C. W.; Levitt, S. H.: Immunotherapy with neuraminidase-treated tumour cells after radiotherapy. Radiat. Res. 64: 485–491 (1975).

396 Sparks, F. C.; Breeding, J. H.: Tumor regression and enhancement resulting from immunotherapy with bacillus Calmette-Guérin and neuraminidase. Cancer Res. 34: 3262–3269 (1974).

397 Spence, R. J.; Simon, R. M.; Baker, A. R.: Failure of immunotherapy with neuraminidase-treated tumor cell vaccine in mice bearing established 3-methylcholanthrene-induced sarcomas. J. natn. Cancer Inst. 60: 451–459 (1978).

398 Spreafico, F.; Mantovani, A.: Immunomodulation by cancer chemotherapeutic agents and antineoplastic activity. Pathobiol. Annu. 11: 177–195 (1981).

399 Springer, G. F.; Desai, P. R.; Banatwala, I.: Blood group MN specific substances and precursors in normal and malignant human breast tissues. Naturwissenschaften 61: 457–458 (1974).

400 Springer, G. F.; Desai, P. R.: Depression of Thomsen-Friedenreich (anti-T) antibody in humans with breast carcinoma. Naturwissenschaften 62: 302–303 (1975).
401 Springer, G. F.; Desai, P. R.: Human blood group MN and precursor specificities: Structural and biological aspects. Carbohydr. Res. 40: 183–192 (1975).
402 Springer, G. F.; Desai, P. R.: Increase in anti-T titer scores of breast carcinoma patients following mastectomy. Naturwissenschaften 62: 587 (1975).
403 Springer, G. F.; Desai, P. R.; Banatwala, I.: Brief communication: Blood group MN antigens and precursors in normal and malignant human breast glandular tissue. J. natn. Cancer Inst. 54: 335–339 (1975).
404 Springer, G. F.; Desai, P. R.; Yang, H. J.; Schachter, H.; Narasimhan, S.: Interrelations of blood group M and precursor specificities and their significance in human carcinoma; in Mohn, Plunkett, Cunningham, Lambert, Human blood groups, p. 179 (Karger, Basel 1977).
405 Springer, G. F.; Desai, P. R.; Murpty, M. S.; Scanlon, E. F.: Delayed-type skin hypersensitivity reaction (DTH) to Thomsen-Friedenreich (T) antigen as diagnostic test for human breast adenocarcinoma. Klin. Wschr. 57: 961–963 (1979).
406 Springer, G. F.; Desai, P. R.; Murthy, M. S.; Tegtmeyer, H.; Scanlon, E. F.: Human carcinoma-associated precursor antigens of the blood group MN system and the host's immune responses to them. Prog. Allergy 26: 42–96 (1979).
407 Springer, G. F.; Desai, P. R.; Murthy, M. S.; Yang, H. J.; Scanlon, E. F.: Precursors of the blood group MN antigens as human carcinoma-associated antigens. Transfusion 19: 233–249 (1979).
408 Springer, G. F.; Murthy, S. M.; Desai, P. R.; Fry, W. A.; Tegtmeyer, J.; Scanlon, E. F.: Patients' immune response to breast and lung carcinoma associated Thomsen-Friedenreich (T) specificity. Klin. Wschr. 60: 121–131 (1982).
409 Springer, G. F.; Desai, P. R.; Fry, W. A.; Goodale, R. L.; Shearen, J. G.; Scanlon, E. F.: T antigen, a tumor marker against which breast, lung, pancreas carcinoma patients mount immune responses. Cancer Detect. Prevent. 6: 111–118 (1983).
410 Staerk, J.; Ronneberger, H. J., Wiegandt, H., Ziegler, W.: Interaction of ganglioside GGtet-1 and its derivatives with choleragen. Eur. J. Biochem. 48: 103–110 (1974).
411 Stewart, G. L.; Parkman, P. D.; Hopps, H. E.; Douglas, R. D.; Hamilton, J. P.; Meyer, H. M. Jr.: Rubella virus hemagglutination inhibition test. New. Engl. J. Med. 276: 554–557 (1967).
412 Sundaresan, P.; Salinis, K. B.; Sundaram, K.; Phondke, G. P.: Differential concentration of surface sialic acid in human T & B lymphocytes. Indian J. exp. Biol. 13: 523–526 (1975).
413 Sur, P.; Roy, D. K.: Antigenicity of Ehrlich ascites through neuraminidase treatment of cells. Indian J. exp. Biol. 17: 953–955 (1979).
414 Suzuki, H.; Kurita, T.; Kakinuma, K.: Effects of neuraminidase on O_2 and H_2O_2 from phagocytosis in human polymorphonuclear leukocytes. Blood 60: 446–453 (1982).
415 Szendröi, Z.; Balogh, F.: Der Prostata-Krebs (Akadémiai Kiadó, Budapest 1965).
416 Takita, H.; Takada, M.; Minowada, J.; Han, T.; Edgerton, F.: Adjuvant immunotherapy of stage III lung carcinoma; in Terry, Windhorst, Immunotherapy of cancer research and therapy, vol. 6, pp. 217–223 (Raven Press, New York 1978).
417 Tao, T. W.; Burger, M. M.: Non-metastasising variants selected from metastasising melanoma cells. Nature, Lond. 270: 437–438 (1977).

418 Tauber, A. I.; Babior, B. M.: Evidence for hydroxyl radical production by human neutrophils. J. clin. Invest. 60: 374–379 (1977).
419 Terry, W. D.; Hodes, R. J.; Rosenberg, S. A.; Fisher, R. I.; Makuch, R.; Gordon, H. G.; Fisher, S. G.: Treatment of stage I and II malignant melanoma with adjuvant immunotherapy of chemotherapy: Preliminary analysis of a prospective, randomized trial; in Terry, Rosenberg, Immunotherapy of human cancer, pp. 251–257 (Elsevier/North Holland, Amsterdam 1980).
420 Thomas, D. B.; Winzler, R. J.: Structural studies on human erythrocyte glycoproteins. Alkali-labile oligosaccharides. J. biol. Chem. 224: 5943–5946 (1969).
421 Thomsen, O.: Ein vermehrungsfähiges Agens als Veränderer des isoagglutinatorischen Verhaltens der roten Blutkörperchen, eine bisher unbekannte Quelle der Fehlbestimmungen. Z. Immunitätsforsch. 52: 85–107 (1927).
422 Treves, A. J.; Carnaud, C.; Trainin, N.: Enhancing T lymphocytes from tumour-bearing mice suppress host resistance to a syngeneic tumour. Eur. J. Immunol. 4: 722–727 (1974).
423 Tsan, M. F.; McIntyre, P. A.: The requirement for membrane sialic acid in the stimulation of superoxide production during phagocytosis by human polymorphonuclear leukocytes. J. exp. Med. 143: 1308–1316 (1976).
424 Tsan, M. F.; Douglass, K. H.; McIntyre, P. A.: Hydrogen peroxide production and killing of Staphylococcus aureus by human polymorphonuclear leukocytes. Blood 49: 437–444 (1977).
425 Tuppy, H.: Preparation, purification, activity and specificity of neuraminidase. Behring Inst. Mitt. 55: 97–103 (1974).
426 Turianskyi, F. H.; Gyenes, L.: The effect of neuraminidase on the sensitivity of tumor cells towards lysis by antibody and complement or by sensitized lymphocytes. Transplantation 22: 24–30 (1976).
427 Turk, J. L.; Parker, D.; Poulter, L. W.: Functional aspects of the selective depletion of lymphoid tissue by cyclophosphamide. Immunology 23: 493–501 (1972).
428 Turk, J. L.; Parker, D.: Further studies on B-lymphocytes suppression on delayed hypersensitivity indicating a possible mechanism for Jones-Mote hypersensitivity. Immunology 24: 751–758 (1973).
429 Turk, J. L.; Parker, D.: Effect of cyclophosphamide on immunological control mechanisms. Immunol. Rev. 65: 99–113 (1982).
430 Überreiter, O.: Der Einfluß von Trächtigkeit und Scheinträchtigkeit auf die Entstehung von Mammatumoren bei der Hündin. Berl. Münchn. tierärztl. Wschr. 23: 451–456 (1966).
431 Uhlenbruck, G.: The Thomsen-Friedenreich (TF) receptor: an old history with new mystery. Immunol. Comm. 10: 251–264 (1981).
432 Uhlenbruck, G.: Metastasenhemmung durch Lektin-Besetzung? Ärztl. Praxis 34: 1723 (1982).
433 Uhlenbruck, G.; Paroe, G. I.; Bird, G. W. G.: On the specificity of lectins with a broad agglutination spectrum. II. Studies on the nature of the T-antigen and the specific receptors for the lectin or Arachis hypogoea (Ground Nut). Z. Immunitätsforsch. 138: 423–433 (1969).
434 Uhlenbruck, G.; Rothe, A.: Biological alterations of membrane-bound glycoproteins by neuraminidase treatment. Behring Inst. Mitt. 55: 177–185 (1974).

435 Ulrichs, K.; Yu, M.-Y.; Duncker, D.; Müller-Ruchholtz, W.: Immunosuppression by cytostatic drugs? Behring Inst. Mitt. *74:* 239–245 (1984).
436 Unanue, E. R.: Secretory function of mononuclear phagocytes. Am. J. Path. *83:* 396–417 (1976).
437 Urbanitz, D.; Büchner, T.; Pielken, H.; Van de Loo, J.: Immunotherapy in the treatment of acute myelogenous leukemia (AML): Rationale, results and future prospects. Klin. Wschr. *61:* 947–954 (1983).
438 Urbanitz, D.; Pielken, H. J.; Koch, P.; Buechner, T.; Hiddemann, W.; Heinecke, A.; Wendt, F.; Maschmeier, G.; Van de Loo, J.: Immunotherapy with allogeneic neuraminidase-treated blasts for maintenance in acute myelogenous leukemia: significant prolongation of remission duration in patients receiving at least 3 cycles of therapy. Onkologie *8:* 157–159 (1985).
439 Valth, P., Uhlenbruck, G.: The Thomsen agglutination phenomenon: a discovery revisited 50 years later. Z. Immunitätsforsch. *154:* 1–15 (1978).
440 Van de Velde, C. J. H.; Van Putten, L. M.; Zwaveling, A.: Effects of regional lymphadenectomy and adjuvant chemotherapy on metastasis and survival in rodent tumour models. Eur. J. Cancer *13:* 883–895 (1977).
441 Van Heyningen, W. E.: Gangliosides as membrane receptors for cholera toxin and serotonin. Nature, Lond. *249:* 415–417 (1974).
442 Varghese, J. N.; Laver, W. G.; Colman, P. M.: Structure of the influenza virus glycoprotein antigen neuraminidase at 2.9.Å resolution. Nature, Lond. *303:* 35–40 (1983).
443 Von Nicolai, H.; Zilliken, F.: Neuraminidase from bifidobacterium bifidum var. pennsylvanicus. Behring Inst. Mitt. *55:* 78–84 (1974).
444 Wahlin, B.; Perlmann, H.; Perlmann, P.: Analysis by a plaque assay of IgG- or IgM-dependent cytotoxic lymphocytes in human blood. J. exp. Med. *144:* 1375–1380 (1976).
445 Wang, T. J.; Miller, H. C.; Esselman, W. J.: Neuraminidase sensitivity of Thy-1 active glycoconjugates. Mol. Immunol. *17:* 1389–1397 (1980).
446 Watkins, E., Jr.: Neuraminidase accentuation of cancer cell immunogenicity. Behring Inst. Mitt. *55:* 355–370 (1974).
447 Watkins, E., Jr.; Ogata, Y.; Anderson, L. L.; Watkins, E., III.; Waters, M. F.: Activation of host lymphocytes cultured with cancer cells treated with neuraminidase. Nature new Biol. *231:* 83–85 (1971).
448 Watkins, E., Jr.; Gray, B. N.; Anderson, L. L.; Baralt, O. L.; Nebril, L. R.; Waters, M. F.; Connery, C. K.: Neuraminidase-mediated augmentation of in vitro immune response of patients with solid tumors. Int. J. Cancer *14:* 799–814 (1974).
449 Weidmann, E.; Sedlacek, H. H.; Lehmann, H. G.; Seiler, F. R.: Klinische Ergebnisse der Tumorimmuntherapie mit Neuraminidase-behandelten Tumorzellen; in König, Immunologie und Tumormarker beim Mammakarzinom, pp. 16–23 (Enke, Stuttgart 1983).
450 Weiner, M. S.; Bianco, C.; Nussenzweig, V.: Enhanced binding of neuraminidase-treated sheep erythrocytes to human T lymphocytes. Blood *42:* 939–946 (1973).
451 Weiss, L.: Studies on cell deformability. I. Effect of surface charge. J. Cell Biol. *26:* 735–739 (1965).
452 Weiss, L.: Studies on cell adhesion in tissue culture. IX. Electrophoretic mobility and contact phenomena. Expl. Cell Res. *51:* 609–625 (1968).

453 Weiss, L.: Studies on cell deformability. V. Some effects of ribunuclease. J. theor. Biol. *18:* 9–18 (1968).
454 Weiss, L.: Neuraminidase, sialic acids and cell interactions. J. natn. Cancer Inst. *50:* 3–19 (1973).
455 Weiss, L.: The topography of cell surface sialic acids and their possible relationship to specific cell interactions. Behring Inst. Mitt. *55:* 185–193 (1974).
456 Werkmeister, J. A.; Pross, H. F.; Roder, J. C.: Modulation of K562 cells with sodium butyrate. Association of impaired NK susceptibility with sialic acid and analysis of other parameters. Int. J. Cancer *32:* 71–78 (1983).
457 Wilson, R. E.; Sonis, S. T.; Godrick, E. A.: Neuraminidase as an adjunct in the treatment of residual systemic tumor with specific immune therapy. Behring Inst. Mitt. *55:* 334–342 (1974).
458 Winchester, R. J.; Man Fu, S.; Winfield, J. B.; Kunkel, H. G.: Immunofluorescent studies on antibodies directed to a buried membrane structure present in lymphocytes and erythrocytes. J. Immunol. *114:* 410–414 (1975).
459 Winfield, J. B.; Lobo, P. I.; Hamilton, M. E.: Fc receptor heterogeneity; immunofluorescent studies of B, T, and 'third population' lymphocytes in human blood with rabbit IgG b4/anti-b4 complexes. J. Immunol. *119:* 1778–1784 (1977).
460 Winzler, R. J.: Carbohydrates in cell surfaces. Int. Rev. Cytol. *29:* 77–125 (1970).
461 Wunderlich, J. R.; Martin, W. J.; Fletcher, F.: Enhanced immunogenicity of syngeneic tumor cells coated with concanavalin A. Fed. Proc. *30:* 246 (1971).
462 Wybran, J.; Carr, M. C.; Fudenberg, H. H.: The human rosette-forming cell as a marker of a population of thymus-derived cells. J. clin. Invest. *51:* 2537–2543 (1972).
463 Wunderlich, M.; Schiessel, R.; Dittrich, C.; Micksche, M.; Sedlacek, H. H.: Effect of adjuvant chemo- or immunotherapy on the prognosis of colorectal cancer operated for cure. Br. J. Surg. *72:* 107–110 (1985).
464 Yogeeswaran, G.; Tao, T. W.: Cell surface sialic acid expression of lectin resistant variant clones of B16 melanoma with altered metastasizing potential. Biochem. Biophys. Res. Commun. *95:* 1452–1460 (1980).
465 Yogeeswaran, G.; Salk, P. L.: Metastatic potential is positively correlated with cell surface sialylation of culture murine tumor cell lines. Science *212:* 1514–1516 (1981).
466 Yu, D. T. Y.: Human lymphocyte receptor movement induced by sheep erythrocyte binding: effect of temperature and neuraminidase treatment. Cell. Immunol. *14:* 313–320 (1974).
467 Zeiller, K.; Schindler, R. K.; Liebich, H.-G.: The T lymphocyte surface in development. A study of the elctrokinetic, antigenic and ultrastructural properties of T lymphocytes in mouse thymus and lymph nodes. Israel J. Scis *11:* 1242–1266 (1975).
468 Zidek, Z.; Capkovà, J.; Bougelik, M.; Masek, K.: Opposite effects of the synthetic immunomodulator, muramyl dipeptide, on rejection of mouse skin allografts. Eur. J. Immunol. *13:* 859–861 (1983).

THE LIBRARY
UNIVERSITY OF CALIFORNIA
San Francisco